This page intentionally left blank

AF873649

This page intentionally left blank

ADVANCED CALCULUS
AND
VECTOR FIELD THEORY

by

KATHLEEN M. URWIN
Department of Mathematics
Battersea College of Technology

PERGAMON PRESS
OXFORD · LONDON · EDINBURGH · NEW YORK
TORONTO · PARIS · BRAUNSCHWEIG

Pergamon Press Ltd., Headington Hill Hall, Oxford
4 & 5 Fitzroy Square, London W.1
Pergamon Press (Scotland) Ltd., 2 & 3 Teviot Place, Edinburgh 1
Pergamon Press Inc., 44-01 21st Street, Long Island City, New York, 11101
Pergamon of Canada Ltd., 6 Adelaide Street East, Toronto, Ontario
Pergamon Press S.A.R.L., 24 rue des Écoles, Paris 5e
Friedr. Vieweg & Sohn Verlag, Postfach 185, 33 Braunschweig, West Germany

---

Copyright © 1966 Pergamon Press Ltd.
First edition 1966
Library of Congress Catalog Card No. 65-25340

Printed in Great Britain by W. & G. Baird Ltd., Belfast

This book is sold subject to the condition that it shall not, by way of trade, be lent, resold, hired out or otherwise disposed of without the publisher's consent, in any form of binding or cover other than that in which it is published.

(1778/66)

CONTENTS

FOREWORD

IN RECENT years, a phenomenal increase has taken place in the amount of mathematical knowledge required by practising engineers and scientists, which in turn demands an unprecedented widening and deepening of their mathematical education. Topics which, only a decade ago, were considered too advanced and abstruse for any but mathematical specialists now form optional parts, at least, of the undergraduate curriculum at many Universities. There is every reason to think that this trend will continue, so that before long all graduate scientists and engineers will be expected to have, not as a luxury but as an essential part of their mathematical equipment, some knowledge of the material which we have sought to cover in the five volumes on "Higher Mathematics for Scientists and Engineers" in the Commonwealth Library.

In broad terms, these are written for the student who has already completed a leisurely two-year course, or a concentrated one-year course, of non-specialist mathematics at University level, reaching a standard which in terms of examinations is approximately that of the Ancillary Mathematics for students of Physics and Chemistry, or the B.Sc.(Eng.) (Part II) of London University. The two volumes *Elementary Mathematics for Scientists and Engineers*, M. D. Hatton, and *Further Mathematics for Scientists and Engineers*, R. Tearse, in the Commonwealth Library series, cover this preliminary material completely, and the five volumes in the Higher Mathematics set rest upon them as their foundation.

To determine the scope of these volumes, and the distribution of material between them, my colleagues and I have taken as our guide the (optional) mathematics section in the syllabus for the

B.Sc. (Engineering) (Part III) degree of London University. The five books in this set cover, respectively, all the material in sections I–V of that examination. We have not, however, hesitated to go some distance beyond these limits from time to time in order to include other topics of special importance, and we hope that as a result the books will meet the needs of most non-mathematical students the world over. This has been particularly the case in the volume on *Numerical Analysis*, where in anticipation of future needs much greater stress has been laid on the use of digital computers than the examination requirements alone would justify. At the same time, we have excluded (sometimes very reluctantly) any major matter which seems unlikely to form, in the foreseeable future, part of an undergraduate course for non-mathematicians.

Although the five volumes may be used independently, they have been written in collaboration and are intended to be complementary, Most of the material on differential equations, for instance, is to be found in *Differential Equations and Special Functions*, but the techniques of numerical solution and the integral–transform method are dealt with respectively in *Numerical Analysis* and *Complex Variable and Integral Transforms*, while some of the underlying calculus is covered in *Advanced Calculus and Vector Field Theory*.

Intended as they are for scientists and engineers, the mathematical presentation of material in these volumes has been deliberately kept at a somewhat unsophisticated level. Nevertheless, we have sought to maintain a reasonable standard of mathematical rigour, and we hope that even the specialist mathematician may find the books useful as introductory material before passing on to stronger meat.

The thanks of the authors are due to members of the Board of Editors of the Engineering division of the Commonwealth Library for many helpful comments and criticisms, both in the planning stages and subsequently; other acknowledgements are in the individual volumes. Collectively, we hope that this project will do something to fill the gap between the resources of mathematics and the needs of natural science and engineering, which is already large and appears to be growing with alarming speed.

F. M. Arscott

Department of Mathematics,
Battersea College of Technology,
*London, S.W.*11

This page intentionally left blank

PREFACE

THIS book falls naturally into two parts. In Chapters 1–5 the basic ideas and techniques of partial differentiation, and of line, multiple and surface integrals are discussed. Chapters 6 and 7 give the elements of vector field theory, taking the integral definitions of the divergence and curl of a vector field as their starting points; the last chapter surveys very briefly some of the immediate applications of vector field theory to five branches of applied mathematics.

Throughout I have given numerous worked examples. In these I have paid particular attention to those points which in my own experience I have found to give most difficulty to students.

In the *text* I have denoted spherical polar coordinates by (r, θ, ψ), and cylindrical polar coordinates by (ρ, ψ, z), so that ψ measures the same angle in both systems. Since there is no one standard notation for these systems, the reader will meet different notations in the course of his reading, and in quoting examination questions in the *exercises* I have kept to the notation of the originals.

The Exercises at the end of each section are intended to give practice in the basic techniques just discussed. The Miscellaneous Exercises are more varied, and contain many examination questions. To give the reader some idea of the standard of a particular problem, I have indicated those taken from *final year* papers. I have classified them under three headings, namely B.Sc.(Eng.), B.Sc.(Spec.) and B.Sc.(Gen.), corresponding to Honours degrees in Engineering and Mathematics, and to the mathematical section of a General Science degree respectively. I should like to thank the Senate of the University of London, and the University of Wales, for permission to use examination questions.

I am very grateful to my colleagues at Battersea for help in checking the answers to examples. I should like to thank particularly Dr. J. H. Wilkinson, who has read the text with great care and interest and made many valuable suggestions, and Prof. F. M. Arscott for his constant help and encouragement throughout the preparation of this volume.

KATHLEEN M. URWIN

Department of Mathematics,
Battersea College of Technology,
*London, S.W.*11

CHAPTER 1

INTRODUCTION

1:1 Introduction

The purpose of this short chapter is to summarise the basic results which are needed for the development of our subject, and with which the reader is assumed to be already familiar. (For a full treatment of these basic topics, see R. Tearse, *Further Mathematics for Engineers and Scientists*, Commonwealth Library Series, Pergamon Press.

1:2 Partial differentiation

Given a function $z(x,y)$ of two independent variables x and y, we can consider the rate of change of z when x is allowed to vary but y is kept constant. This rate of change is the partial derivative of z with respect to x and is written $\partial z/\partial x$ or $z_x(x,y)$, or briefly, z_x. Similarly, we obtain $\partial z/\partial y$, the partial derivative of z with respect to y, x being now constant. In practice, $\partial z/\partial x$ is found by differentiating z with respect to x by the ordinary processes of calculus, treating y as a constant. For example, if

$$z = 3x^3 + 2x^2y + y^3,$$

then

$$\frac{\partial z}{\partial x} = 9x^2 + 4xy, \quad \frac{\partial z}{\partial y} = 2x^2 + 3y^2.$$

In general $\partial z/\partial x$ is a function of both x and y; so we can form its partial derivatives. We thus get two second order derivatives, namely

$$\frac{\partial}{\partial x}\left(\frac{\partial z}{\partial x}\right) \quad \text{and} \quad \frac{\partial}{\partial y}\left(\frac{\partial z}{\partial x}\right).$$

These are written as $\partial^2 z/\partial x^2$, or z_{xx}, and $\partial^2 z/\partial y\,\partial x$, or z_{yx}. Further, we obtain similarly two second order derivatives from $\partial z/\partial y$. These are $\partial^2 z/\partial x\,\partial y$ and $\partial^2 z/\partial y^2$. These second order derivatives are in general functions of both x and y, and hence themselves possess partial derivatives. In the example above,

$$\frac{\partial^2 z}{\partial x^2} = 18x + 4y, \quad \frac{\partial^2 z}{\partial y\,\partial x} = \frac{\partial}{\partial y}(9x^2 + 4xy) = 4x,$$

$$\frac{\partial^2 z}{\partial y^2} = 6y, \quad \frac{\partial^2 z}{\partial x\,\partial y} = \frac{\partial}{\partial x}(2x^2 + 3y^2) = 4x.$$

We note that here $\partial^2 z/\partial y\,\partial x = \partial^2 z/\partial x\,\partial y$. In fact, provided $z(x,y)$ obeys certain continuity conditions, then $\partial^2 z/\partial y\,\partial x$ is always the same as $\partial^2 z/\partial x\,\partial y$. We shall assume this property throughout the following pages. (For a discussion of this result, see R. P. Gillespie, *Partial Differentiation*, Oliver & Boyd, 1939.)

We can form derivatives such as

$$\frac{\partial^3 z}{\partial x^3}, \quad \frac{\partial^3 z}{\partial x^2\,\partial y}\left[= \frac{\partial}{\partial x}\left(\frac{\partial^2 z}{\partial x\,\partial y}\right)\right], \quad \frac{\partial^3 z}{\partial x\,\partial y\,\partial x},$$

and so on. We note that

$$\frac{\partial^3 z}{\partial x^2\,\partial y} = \frac{\partial^3 z}{\partial x\,\partial y\,\partial x}, \quad \text{since} \quad \frac{\partial^2 z}{\partial x\,\partial y} = \frac{\partial^2 z}{\partial y\,\partial x}.$$

A function of n independent variables possesses n first order derivatives, which we obtain by normal differentiation with respect to one particular variable, in turn, treating all the others as constants. Thus, if

$$z = x_1^2 + x_2^2 + \cdots + x_n^2, \quad \frac{\partial z}{\partial x_1} = 2x_1, \quad \frac{\partial z}{\partial x_3} = 2x_3, \text{ etc.}$$

We shall be concerned mainly with functions of two or three variables.

1:3 Taylor's Theorem

For a function $f(x)$ of one variable, Taylor's Theorem may be stated as

$$f(a+h) = f(a) + hf'(a) + \frac{1}{2!}h^2 f''(a) + \cdots + \frac{1}{n!}h^n f^n(a) + R_n,$$

where R_n is the remainder, which is a function of h and a, and may be expressed in various standard forms. $f(x)$ must be such that its first $(n+1)$ derivatives are continuous for $a \leqq x \leqq a+h$.

The corresponding result for functions of two independent variables is

$$f(a+h, b+k) = f(a,b) + \left(h\frac{\partial}{\partial x} + k\frac{\partial}{\partial y}\right)f(a,b) + \frac{1}{2!}\left(h\frac{\partial}{\partial x} + k\frac{\partial}{\partial y}\right)^2 f(a,b) + \cdots + \frac{1}{n!}\left(h\frac{\partial}{\partial x} + k\frac{\partial}{\partial y}\right)^n f(a,b) + R_n,$$

where

$$\left(h\frac{\partial}{\partial x} + k\frac{\partial}{\partial y}\right)^2 f$$

is a symbolic way of saying that the operator

$$\left(h\frac{\partial}{\partial x} + k\frac{\partial}{\partial y}\right)$$

is applied twice to f. Thus

$$\left(h\frac{\partial}{\partial x} + k\frac{\partial}{\partial y}\right)^2 f = \left(h\frac{\partial}{\partial x} + k\frac{\partial}{\partial y}\right)\left(h\frac{\partial f}{\partial x} + k\frac{\partial f}{\partial y}\right) = h^2\frac{\partial^2 f}{\partial x^2} + 2hk\frac{\partial^2 f}{\partial x\,\partial y} + k^2\frac{\partial^2 f}{\partial y^2},$$

$$\left(h\frac{\partial}{\partial x} + k\frac{\partial}{\partial y}\right)^3 f = \left(h\frac{\partial}{\partial x} + k\frac{\partial}{\partial y}\right)\left(h\frac{\partial}{\partial x} + k\frac{\partial}{\partial y}\right)^2 f, \text{ etc.,}$$

and x, y are replaced by a, b after differentiation.

The continuity conditions may be stated in various ways; one useful form is that $f(x, y)$ and *all* its partial derivatives of orders up to and including $(n + 1)$ are continuous.

The theorem may be extended to functions of any number of variables.

1:4 Vector algebra

The vector $\mathbf{a}$ is in the direction of the unit vector $\hat{\mathbf{a}}$ and of magnitude a. So $\mathbf{a} = a\hat{\mathbf{a}}$. We often write $a = |\mathbf{a}|$. If $\mathbf{a}$ is expressed in terms of three mutually orthogonal unit vectors $\mathbf{i}$, $\mathbf{j}$, $\mathbf{k}$ as $\mathbf{a} = a_1\mathbf{i} + a_2\mathbf{j} + a_3\mathbf{k}$, where a_1, a_2, a_3 are scalar quantities, then

$$|\mathbf{a}| = a = (a_1^2 + a_2^2 + a_3^2)^{\frac{1}{2}}. \tag{1}$$

The vector $o\mathbf{i} + o\mathbf{j} + o\mathbf{k}$ is the *null* or *zero* vector and is denoted by $\mathbf{0}$.

The scalar product of two vectors $\mathbf{a}$ and $\mathbf{b}$ is defined as $ab\cos\theta$, where θ is the angle between $\mathbf{a}$ and $\mathbf{b}$, and $0 \leqq \theta \leqq \pi$. This product is written $\mathbf{a} \,.\, \mathbf{b}$ and we note that $\mathbf{a} \,.\, \mathbf{b} = \mathbf{b} \,.\, \mathbf{a}$.
If $\mathbf{a} = a_1\mathbf{i} + a_2\mathbf{j} + a_3\mathbf{k}$, $\mathbf{b} = b_1\mathbf{i} + b_2\mathbf{j} + b_3\mathbf{k}$, then

$$\mathbf{a} \,.\, \mathbf{b} = a_1b_1 + a_2b_2 + a_3b_3. \tag{2}$$

If $\theta = 0$, $\mathbf{a} \,.\, \mathbf{b} = ab$ and if $\mathbf{a} = \mathbf{b}$, $\mathbf{a} \,.\, \mathbf{a} = a^2$. Note that $\mathbf{a} \,.\, \mathbf{a}$ is often written $\mathbf{a}^2$. If $\theta = \frac{1}{2}\pi$, $\mathbf{a} \,.\, \mathbf{b} = 0$.

The vector product of two vectors $\mathbf{a}$ and $\mathbf{b}$ is defined as the vector $(ab\sin\theta)\hat{\mathbf{n}}$, $0 \leqq \theta \leqq \pi$, where $\hat{\mathbf{n}}$ is a unit vector perpendicular to the plane of $\mathbf{a}$ and $\mathbf{b}$, and such that $\mathbf{a}$, $\mathbf{b}$, $\hat{\mathbf{n}}$ form a right-handed set. This product is written $\mathbf{a} \times \mathbf{b}$, or sometimes $\mathbf{a} \wedge \mathbf{b}$, and we note from the definition that $\mathbf{a} \times \mathbf{b} = -\mathbf{b} \times \mathbf{a}$. In component form

$$\mathbf{a} \times \mathbf{b} = (a_2b_3 - a_3b_2)\mathbf{i} + (a_3b_1 - a_1b_3)\mathbf{j} + (a_1b_2 - a_2b_1)\mathbf{k}. \tag{3}$$

A most useful form for this product is given by the determinant

$$\begin{vmatrix} \mathbf{i} & \mathbf{j} & \mathbf{k} \\ a_1 & a_2 & a_3 \\ b_1 & b_2 & b_3 \end{vmatrix}. \tag{4}$$

We can form the scalar product of $\mathbf{a} \times \mathbf{b}$ with $\mathbf{c}$. This is the scalar triple product of $\mathbf{a}$, $\mathbf{b}$ and $\mathbf{c}$. We write this $\mathbf{a} \times \mathbf{b} \,.\, \mathbf{c}$ or $[\mathbf{a}, \mathbf{b}, \mathbf{c}]$. The standard results are

$$[\mathbf{a}, \mathbf{b}, \mathbf{c}] = [\mathbf{b}, \mathbf{c}, \mathbf{a}] = [\mathbf{c}, \mathbf{a}, \mathbf{b}]$$
$$= -[\mathbf{b}, \mathbf{a}, \mathbf{c}] = -[\mathbf{c}, \mathbf{b}, \mathbf{a}] = -[\mathbf{a}, \mathbf{c}, \mathbf{b}].$$

If $\mathbf{c} = c_1\mathbf{i} + c_2\mathbf{j} + c_3\mathbf{k}$, then, in component form,

$$[\mathbf{a}, \mathbf{b}, \mathbf{c}] = \begin{vmatrix} a_1 & a_2 & a_3 \\ b_1 & b_2 & b_3 \\ c_1 & c_2 & c_3 \end{vmatrix}.$$

We can also form the vector product of $\mathbf{a}$ with $\mathbf{b} \times \mathbf{c}$. This is the vector triple product of $\mathbf{a}$, $\mathbf{b}$ and $\mathbf{c}$. We write it $\mathbf{a} \times (\mathbf{b} \times \mathbf{c})$, and note that $\mathbf{a} \times (\mathbf{b} \times \mathbf{c}) = -(\mathbf{b} \times \mathbf{c}) \times \mathbf{a}$. The standard result is

$$\mathbf{a} \times (\mathbf{b} \times \mathbf{c}) = (\mathbf{a} \,.\, \mathbf{c})\mathbf{b} - (\mathbf{a} \,.\, \mathbf{b})\mathbf{c}.$$

In Chapters 6 and 7 we shall always use the vector $\mathbf{r}$ to denote the position vector of the point $P(x, y, z)$ referred to cartesian axes $Oxyz$. Thus

$$\overrightarrow{OP} = \mathbf{r} = x\mathbf{i} + y\mathbf{j} + z\mathbf{k} \quad \text{and} \quad r^2 = x^2 + y^2 + z^2.$$

The miscellaneous examples which follow are intended to supply brief revision in elementary partial differentiation and in vector algebra. It is essential that the reader should be able to evaluate and use the scalar and vector products of vectors accurately before attempting to read Vector Field Theory.

Miscellaneous Exercises 1

1. If $z = (x + y) \log (y/x)$, show that

$$\text{(i)}\ x\frac{\partial z}{\partial x} + y\frac{\partial z}{\partial y} = z, \qquad \text{(ii)}\ \frac{\partial^2 z}{\partial x^2} \cdot \frac{\partial^2 z}{\partial y^2} = \left(\frac{\partial^2 z}{\partial x\, \partial y}\right)^2.$$

2. If $z = e^{y-x} \sin (y - x)$, show that

$$\text{(i)}\ \frac{\partial^2 z}{\partial x^2} = \frac{\partial^2 z}{\partial y^2}, \qquad \text{(ii)}\ \left(\frac{\partial^3 z}{\partial x^3}\right)^2 - \frac{\partial^2 z}{\partial x^2} \cdot \frac{\partial^4 z}{\partial x^4} = 4\,e^{2(y-x)}.$$

3. If $V = \log \sqrt{\{(x-a)^2 + (y-b)^2 + (z-c)^2\}}$, evaluate
(i) $V_{xx} + V_{yy} + V_{zz}$, (ii) $V_{xy} + V_{yz} + V_{zx}$.

4. If $V = e^x \sin y + e^y \sin x$, evaluate

(i) $\dfrac{\partial^2 V}{\partial x^2} + \dfrac{\partial^2 V}{\partial y^2}$, (ii) $\left(\dfrac{\partial^2 V}{\partial x \partial y}\right)^2 - \dfrac{\partial^2 V}{\partial x^2} \cdot \dfrac{\partial^2 V}{\partial y^2}$,

(iii) $\left(\dfrac{\partial^2 V}{\partial x^2} - \dfrac{\partial^2 V}{\partial y^2}\right)^2 + 4\left\{\left(\dfrac{\partial^2 V}{\partial x \partial y}\right)^2 - \left(\dfrac{\partial^3 V}{\partial y \partial x^2}\right)^2 - \left(\dfrac{\partial^3 V}{\partial x \partial y^2}\right)^2\right\}$.

5. Evaluate (i) $(\mathbf{i} + 3\mathbf{j} + 2\mathbf{k}) . (4\mathbf{i} - \mathbf{j} + \mathbf{k})$,
(ii) $(\mathbf{i} + 3\mathbf{j} + 2\mathbf{k}) \times (4\mathbf{i} - \mathbf{j} + \mathbf{k})$, (iii) $3\mathbf{j} \times (\mathbf{k} + \mathbf{i})$,
(iv) $(3\mathbf{j} + 4\mathbf{k}) . (3\mathbf{k} + 4\mathbf{i})$.

6. If $\mathbf{A} = \mathbf{i} + 3\mathbf{j} + 2\mathbf{k}$, $\mathbf{B} = 4\mathbf{i} - \mathbf{j} + \mathbf{k}$, evaluate
(i) $|\mathbf{A}|$, (ii) $|\mathbf{B}|$, (iii) $|\mathbf{A} \times \mathbf{B}|$.
Hence write down the unit vectors $\hat{\mathbf{A}}$, $\hat{\mathbf{B}}$, $\hat{\mathbf{C}}$, in the directions given by the vectors $\mathbf{A}$, $\mathbf{B}$, $\mathbf{A} \times \mathbf{B}$ respectively.

7. Two vectors are given by $\mathbf{A} = 2\mathbf{i} + 3\mathbf{j} + 4\mathbf{k}$, $\mathbf{B} = 5\mathbf{i} - 2\mathbf{j} - \mathbf{k}$.
Evaluate $\mathbf{A} \times \mathbf{B}$ and verify that $(\mathbf{A} - \mathbf{B}) \times (\mathbf{A} + \mathbf{B}) = 2(\mathbf{A} \times \mathbf{B})$.

8. If **a**, **b** and **c** are three vectors, show, by expressing them in cartesian coordinates, or otherwise, that

$$\mathbf{a} \times (\mathbf{b} \times \mathbf{c}) = (\mathbf{a} . \mathbf{c})\mathbf{b} - (\mathbf{a} . \mathbf{b})\mathbf{c}.$$

Use this result to find the vector **x** which satisfies the equations

$$\mathbf{x} \times \mathbf{a} = \mathbf{b}, \quad \mathbf{x} . \mathbf{c} = p,$$

in which p is a given scalar and $\mathbf{a} . \mathbf{c} \neq 0$.

9. Establish the identities
(i) $(\mathbf{a} \times \mathbf{b}) . (\mathbf{c} \times \mathbf{d}) = (\mathbf{a} . \mathbf{c})(\mathbf{b} . \mathbf{d}) - (\mathbf{b} . \mathbf{c})(\mathbf{a} . \mathbf{d})$,
(ii) $(\mathbf{a} \times \mathbf{b}) \times (\mathbf{c} \times \mathbf{d}) = [\mathbf{a}, \mathbf{b}, \mathbf{d}]\mathbf{c} - [\mathbf{a}, \mathbf{b}, \mathbf{c}]\mathbf{d}$,
(iii) $[\mathbf{a} \times \mathbf{b}, \mathbf{a} \times \mathbf{c}, \mathbf{d}] = (\mathbf{a} . \mathbf{d})[\mathbf{a}, \mathbf{b}, \mathbf{c}]$,
where **a**, **b**, **c**, **d** are any four vectors.

10. Interpret the triple scalar product [**a**, **b**, **c**] geometrically, and find its value in terms of the cartesian components of **a**, **b**, **c**.
Prove directly that if **F** and **r** are two vectors such that $\mathbf{r} . \mathbf{F} = 0$, then $\mathbf{F} \times (\mathbf{r} \times \mathbf{F}) = F^2\mathbf{r}$. If the above vectors **F** and **r**, together with a third vector **G**, satisfy the equation

$$\mathbf{G} - (\mathbf{r} \times \mathbf{F}) = \lambda \mathbf{F},$$

where λ is a scalar, show by taking a dot and cross product of the equation with a suitable vector that

(i) $\lambda = \dfrac{\mathbf{F} . \mathbf{G}}{F^2}$, (ii) $\mathbf{r} = \dfrac{\mathbf{F} \times \mathbf{G}}{F^2}$. [B.Sc.(Eng.)]

11. If $\mathbf{a}_1$, $\mathbf{a}_2$, $\mathbf{a}_3$, $\mathbf{a}_4$ are any four vectors, while $\mathbf{b}_1$, $\mathbf{b}_2$, $\mathbf{b}_3$, $\mathbf{b}_4$ are defined by

$$\mathbf{b}_1 = \mathbf{a}_2 \times \mathbf{a}_3 + \mathbf{a}_3 \times \mathbf{a}_4 + \mathbf{a}_4 \times \mathbf{a}_2,$$

and the equations formed from this by the cyclic interchanges of the suffixes 1, 2, 3, 4, prove that
(i) $\mathbf{b}_1 - \mathbf{b}_2 + \mathbf{b}_3 - \mathbf{b}_4 = \mathbf{0}$,
(ii) $\mathbf{a}_1 \times \mathbf{b}_1 - \mathbf{a}_2 \times \mathbf{b}_2 + \mathbf{a}_3 \times \mathbf{b}_3 - \mathbf{a}_4 \times \mathbf{b}_4 = \mathbf{0}$.

12. Find two vectors $\mathbf{x} = a\mathbf{i} + b\mathbf{j} + c\mathbf{k}$ such that

$$(\mathbf{i} \times \mathbf{x}) \times \mathbf{j} = \mathbf{i}, \quad (\mathbf{j} \times \mathbf{x}) \times \mathbf{i} = \mathbf{j}, \quad \mathbf{x}^2 = 11.$$

CHAPTER 2

FUNCTIONS OF SEVERAL INDEPENDENT VARIABLES

2:1 The Chain Rule

In the following pages we shall be concerned only with *differentiable* functions. Before giving the general definition of such a function of two independent variables, let us consider an example. Suppose that $f(x,y)$ is a function of both x and y, defined by the equation $f(x,y) = x^2 + y^2$. Let x and y be given small increments δx, δy respectively, the corresponding increment in f being δf. Then

$$\begin{aligned}\delta f &= (x + \delta x)^2 + (y + \delta y)^2 - (x^2 + y^2) \\ &= 2x\delta x + 2y\delta y + (\delta x)^2 + (\delta y)^2. \end{aligned} \tag{1}$$

Now $\partial f/\partial x = 2x$, $\partial f/\partial y = 2y$. So (1) can be written

$$\delta f = \frac{\partial f}{\partial x}\delta x + \frac{\partial f}{\partial y}\delta y + (\delta x^2 + \delta y^2), \tag{2}$$

where δx^2 is used to denote $(\delta x)^2$.

Any function $f(x,y)$, which is such that δf is given by an expression of the form

$$\delta f = \frac{\partial f}{\partial x}\delta x + \frac{\partial f}{\partial y}\delta y + \varepsilon\sqrt{(\delta x^2 + \delta y^2)}, \tag{3}$$

where ε is in general a function of δx, δy, x and y such that $\varepsilon \to 0$ as we make $\sqrt{(\delta x^2 + \delta y^2)} \to 0$, is said to be *differentiable*. In the above example, it is clear from (2) that the condition on ε is satisfied for $\varepsilon = \sqrt{(\delta x^2 + \delta y^2)}$, so that $x^2 + y^2$ is a differentiable function of x and y. This definition of differentiability implies that we are dealing with a function $f(x,y)$ for which δf is always made up of two parts:

(i) $\dfrac{\partial f}{\partial x}\delta x + \dfrac{\partial f}{\partial y}\delta y$, which is *linear* in δx and δy,

(ii) the terms which are not linear in δx and δy, but which, taken together, reduce to zero as $\sqrt{(\delta x^2 + \delta y^2)}$ tends to zero, that is as both δx and δy tend to zero simultaneously. Any polynomial function in x and y satisfies this condition; so do the exponential, sine and cosine functions, e.g. $\exp(ax + by)$, $\sin(x + y)$, $\cos(2x - 3y)$, and so on.

The definition extends at once to a function of any number of variables.

Now we suppose that x and y are themselves given as differentiable functions of two other independent variables u and v. Then f is a function of u and v and we require to find $\partial f/\partial u$, $\partial f/\partial v$. Substitution for x and y in terms of u and v is practicable in only the simplest cases. We prove now the fundamental formulae for $\partial f/\partial u$, $\partial f/\partial v$, namely:

$$\frac{\partial f}{\partial u} = \frac{\partial f}{\partial x}\frac{\partial x}{\partial u} + \frac{\partial f}{\partial y}\frac{\partial y}{\partial u}, \tag{4}$$

$$\frac{\partial f}{\partial v} = \frac{\partial f}{\partial x}\frac{\partial x}{\partial v} + \frac{\partial f}{\partial y}\frac{\partial y}{\partial v}. \tag{5}$$

Proof. Suppose that u is given an increment δu whilst v is kept constant. Let δx, δy, δf be the corresponding increments in x, y and f. Since f is a differentiable function of x and y,

$$\delta f = \frac{\partial f}{\partial x}\delta x + \frac{\partial f}{\partial y}\delta y + \varepsilon\sqrt{(\delta x^2 + \delta y^2)},$$

where $\varepsilon \to 0$ as $\sqrt{(\delta x^2 + \delta y^2)} \to 0$. Dividing by δu,

$$\frac{\delta f}{\delta u} = \frac{\partial f}{\partial x}\frac{\delta x}{\delta u} + \frac{\partial f}{\partial y}\frac{\delta y}{\delta u} + \varepsilon\sqrt{\left\{\left(\frac{\delta x}{\delta u}\right)^2 + \left(\frac{\delta y}{\delta u}\right)^2\right\}}.$$

Now let $\delta u \to 0$. Then

$$\frac{\delta x}{\delta u} \to \frac{\partial x}{\partial u}, \quad \frac{\delta y}{\delta u} \to \frac{\partial y}{\partial u}, \quad \frac{\delta f}{\delta u} \to \frac{\partial f}{\partial u}.$$

We are left to show that $\varepsilon \to 0$ in order to obtain the result stated in (4). We know that ε will tend to zero if $\sqrt{(\delta x^2 + \delta y^2)} \to 0$. Here, since

$$\sqrt{(\delta x^2 + \delta y^2)} = \sqrt{\left\{\left(\frac{\delta x}{\delta u}\right)^2 + \left(\frac{\delta y}{\delta u}\right)^2\right\}}\,\delta u,$$

it follows that as $\delta u \to 0$ so does $\sqrt{(\delta x^2 + \delta y^2)}$, and hence so does ε. Thus (4) is established. Similarly, if we allow v to vary and keep u constant, (5) is obtained.

The rule given by (4) and (5) is known as the *Chain Rule* for change of variable. It extends at once to any number of variables. In particular for $f(x,y,z)$, where x, y and z are themselves functions of u, v and w, we get

$$\frac{\partial f}{\partial u} = \frac{\partial f}{\partial x}\frac{\partial x}{\partial u} + \frac{\partial f}{\partial y}\frac{\partial y}{\partial u} + \frac{\partial f}{\partial z}\frac{\partial z}{\partial u}, \tag{6}$$

$$\frac{\partial f}{\partial v} = \frac{\partial f}{\partial x}\frac{\partial x}{\partial v} + \frac{\partial f}{\partial y}\frac{\partial y}{\partial v} + \frac{\partial f}{\partial z}\frac{\partial z}{\partial v}, \tag{7}$$

$$\frac{\partial f}{\partial w} = \frac{\partial f}{\partial x}\frac{\partial x}{\partial w} + \frac{\partial f}{\partial y}\frac{\partial y}{\partial w} + \frac{\partial f}{\partial z}\frac{\partial z}{\partial w}. \tag{8}$$

Various special cases arise. For example, if we have a function $f(x,y)$, where x and y are each functions of one variable t, then equation (4) reduces to

$$\frac{\mathrm{d}f}{\mathrm{d}t} = \frac{\partial f}{\partial x}\frac{\mathrm{d}x}{\mathrm{d}t} + \frac{\partial f}{\partial y}\frac{\mathrm{d}y}{\mathrm{d}t}. \tag{9}$$

Similarly, for a function $f(u)$ where u is a function of x and y. equations (4) and (5) give

$$\frac{\partial f}{\partial x} = \frac{\mathrm{d}f}{\mathrm{d}u}\frac{\partial u}{\partial x}, \quad \frac{\partial f}{\partial y} = \frac{\mathrm{d}f}{\mathrm{d}u}\frac{\partial u}{\partial y}. \tag{10}$$

For example, suppose $z = f(x/y)$. Then

$$\frac{\partial z}{\partial x} = \frac{df}{du}\frac{\partial u}{\partial x}, \quad \frac{\partial z}{\partial y} = \frac{df}{du}\frac{\partial u}{\partial y},$$

where $u = x/y$. So

$$\frac{\partial z}{\partial x} = \frac{1}{y}\frac{df}{du}, \quad \frac{\partial z}{\partial y} = -\frac{x}{y^2}\frac{df}{du}.$$

Exercises 2:1

1. If $z = x^n e^{ax+by} f(y/x)$, where $f(y/x)$ is an arbitrary function of y/x, show that

$$x\frac{\partial z}{\partial x} + y\frac{\partial z}{\partial y} = (ax + by + n)z.$$

2. If V is any function of r and θ, where $r^2 = x^2 + y^2 + z^2$, $z \tan \theta = \sqrt{(x^2 + y^2)}$, show that

$$x\frac{\partial V}{\partial x} + y\frac{\partial V}{\partial y} + z\frac{\partial V}{\partial z} = r\frac{\partial V}{\partial r}.$$

3. If $Z = f(u)F(v)$, where $u = x + y$, $v = y/x$, and both f and F are arbitrary functions, show that

$$x\frac{\partial z}{\partial x} + y\frac{\partial z}{\partial y} = uf'(u)F(v),$$

$f'(u)$ being the derivative of $f(u)$ with respect to u.

4. The function $f(x,y)$ is defined to be *homogeneous* of degree n if $f(tx,ty) = t^n f(x,y)$ for all t. Show that $f(x,y) = (x+y)\log(y/x)$ is homogeneous, and verify that

$$x\frac{\partial f}{\partial x} + y\frac{\partial f}{\partial y} = f(x, y).$$

Show also that if $w = x^3y^2 + z^2yx^2 + z^5 \sin(x/y)$, then

$$x\frac{\partial w}{\partial x} + y\frac{\partial w}{\partial y} + z\frac{\partial w}{\partial z} = 5w.$$

(For a homogeneous function $f(x,y,z)$ of degree n, Euler's theorem states that

$$x\frac{\partial f}{\partial x} + y\frac{\partial f}{\partial y} + z\frac{\partial f}{\partial z} = nf.$$

For a proof, see R. Tearse, *Further Mathematics for Engineers and Scientists*.)

2:2 Change of Variable

Given a function $f(x, y)$, where x and y are both functions of u and v, we need to be able to find the second order derivatives $\partial^2 f/\partial u^2$, $\partial^2 f/\partial u \partial v$, $\partial^2 f/\partial v^2$. Each of these will, in general, be expressed in terms of $\partial f/\partial x$, $\partial f/\partial y$, $\partial^2 f/\partial x^2$, $\partial^2 f/\partial x \partial y$, $\partial^2 f/\partial y^2$.

The method is best illustrated by an example.

EXAMPLE. *If $x = (u - v)^2$, $y = (u + v)^2$, and $f(x, y)$ is any differentiable function of x and y, find expressions for $\partial^2 f/\partial u^2$, $\partial^2 f/\partial v^2$, $\partial^2 f/\partial u \partial v$.*

$$\frac{\partial f}{\partial u} = \frac{\partial f}{\partial x}\frac{\partial x}{\partial u} + \frac{\partial f}{\partial y}\frac{\partial y}{\partial u} = 2(u - v)\frac{\partial f}{\partial x} + 2(u + v)\frac{\partial f}{\partial y}, \tag{1}$$

$$\frac{\partial f}{\partial v} = \frac{\partial f}{\partial x}\frac{\partial x}{\partial v} + \frac{\partial f}{\partial y}\frac{\partial y}{\partial v} = -2(u - v)\frac{\partial f}{\partial x} + 2(u + v)\frac{\partial f}{\partial y}. \tag{2}$$

From equations (1) and (2) we derive the *operator equations*

$$\frac{\partial}{\partial u} = 2(u - v)\frac{\partial}{\partial x} + 2(u + v)\frac{\partial}{\partial y}, \tag{3}$$

$$\frac{\partial}{\partial v} = -2(u - v)\frac{\partial}{\partial x} + 2(u + v)\frac{\partial}{\partial y}. \tag{4}$$

Equation (3) tells us that for the given transformation from x, y to u, v the effect on a function of the *partial differential operator* $\partial/\partial u$ is the same as that of the operator denoted by

$$\left\{2(u - v)\frac{\partial}{\partial x} + 2(u + v)\frac{\partial}{\partial y}\right\}.$$

We may replace $\partial/\partial u$ by its equivalent operator

$$\left\{2(u - v)\frac{\partial}{\partial x} + 2(u + v)\frac{\partial}{\partial y}\right\}$$

at a suitable stage in the process of differentiation. For example, suppose we are finding $\partial^2 f/\partial u^2$. Then

$$\frac{\partial^2 f}{\partial u^2} = \frac{\partial}{\partial u}\left(\frac{\partial f}{\partial u}\right) = \frac{\partial}{\partial u}\left\{2(u-v)\frac{\partial f}{\partial x} + 2(u+v)\frac{\partial f}{\partial y}\right\},$$

so

$$\frac{\partial^2}{\partial u^2} = 2\frac{\partial f}{\partial x} + 2(u-v)\frac{\partial}{\partial u}\left(\frac{\partial f}{\partial x}\right) + 2\frac{\partial f}{\partial y} + 2(u+v)\frac{\partial}{\partial u}\left(\frac{\partial f}{\partial y}\right). \tag{5}$$

In equation (5), no further differentiation with respect to u is possible on the right-hand side as it stands. So we now use equation (3):

$$\frac{\partial^2 f}{\partial u^2} = 2\left(\frac{\partial f}{\partial x} + \frac{\partial f}{\partial y}\right) + 2(u-v)\left\{2(u-v)\frac{\partial}{\partial x} + 2(u+v)\frac{\partial}{\partial y}\right\}\left(\frac{\partial f}{\partial x}\right) +$$

$$+ 2(u+v)\left\{2(u-v)\frac{\partial}{\partial x} + 2(u+v)\frac{\partial}{\partial y}\right\}\left(\frac{\partial f}{\partial y}\right)$$

$$= 2\left(\frac{\partial f}{\partial x} + \frac{\partial f}{\partial y}\right) + 2(u-v)\left\{2(u-v)\frac{\partial^2 f}{\partial x^2} + 2(u+v)\frac{\partial^2 f}{\partial y\,\partial x}\right\} +$$

$$+ 2(u+v)\left\{2(u-v)\frac{\partial^2 f}{\partial x\,\partial y} + 2(u+v)\frac{\partial^2 f}{\partial y^2}\right\}$$

$$= 2\left(\frac{\partial f}{\partial x} + \frac{\partial f}{\partial y}\right) + 4(u-v)^2\frac{\partial^2 f}{\partial x^2} + 8(u^2-v^2)\frac{d^2 f}{\partial x\,\partial y} +$$

$$+ 4(u+v)^2\frac{\partial^2 f}{\partial y^2}$$

Thus

$$\frac{\partial^2 f}{\partial u^2} = 2\frac{\partial f}{\partial x} + 2\frac{\partial f}{\partial y} + 4x\frac{\partial^2 f}{\partial x^2} + 8\sqrt{(xy)}\frac{\partial^2 f}{\partial x\,\partial y} + 4y\frac{\partial^2 f}{\partial y^2}.$$

Similarly, we can find $\partial^2 f/\partial u\,\partial v$; for

$$\frac{\partial^2 f}{\partial u\,\partial v} = \frac{\partial}{\partial u}\left(\frac{\partial f}{\partial v}\right)$$

$$= \frac{\partial}{\partial u}\left\{-2(u-v)\frac{\partial f}{\partial x} + 2(u+v)\frac{\partial f}{\partial y}\right\}$$

$$= -2\frac{\partial f}{\partial x} - 2(u-v)\frac{\partial}{\partial u}\left(\frac{\partial f}{\partial x}\right) + 2\frac{\partial f}{\partial y} + 2(u+v)\frac{\partial}{\partial u}\left(\frac{\partial f}{\partial y}\right) \tag{6}$$

$$= 2\left(\frac{\partial f}{\partial y} - \frac{\partial f}{\partial x}\right) - 2(u-v)\left\{2(u-v)\frac{\partial}{\partial x} + 2(u+v)\frac{\partial}{\partial y}\right\}\left(\frac{\partial f}{\partial x}\right) +$$

$$+ 2(u+v)\left\{2(u-v)\frac{\partial}{\partial x} + 2(u+v)\frac{\partial}{\partial y}\right\}\left(\frac{\partial f}{\partial y}\right)$$

$$= 2\left(\frac{\partial f}{\partial y} - \frac{\partial f}{\partial x}\right) - 4(u-v)^2\frac{\partial^2 f}{\partial x^2} + 4(u+v)^2\frac{\partial^2 f}{\partial y^2}.$$

So,
$$\frac{\partial^2 f}{\partial u\,\partial v} = 2\frac{\partial f}{\partial y} - 2\frac{\partial f}{\partial x} - 4x\frac{\partial^2 f}{\partial x^2} + 4y\frac{\partial^2 f}{\partial y^2}.$$

Finally,

$$\frac{\partial^2 f}{\partial v^2} = \frac{\partial}{\partial v}\left\{-2(u-v)\frac{\partial f}{\partial x} + 2(u+v)\frac{\partial f}{\partial y}\right\}$$

$$= 2\frac{\partial f}{\partial x} + 2\frac{\partial f}{\partial y} - 2(u-v)\frac{\partial}{\partial v}\left(\frac{\partial f}{\partial x}\right) + 2(u+v)\frac{\partial}{\partial v}\left(\frac{\partial f}{\partial y}\right). \tag{7}$$

Now we use equation (4), and replace $\partial/\partial v$ by its equivalent operator:

$$\frac{\partial^2 f}{\partial v^2} = 2\frac{\partial f}{\partial x} + 2\frac{\partial f}{\partial y} - 2(u-v)\left\{-2(u-v)\frac{\partial}{\partial x} + 2(u+v)\frac{\partial}{\partial y}\right\}\left(\frac{\partial f}{\partial x}\right) +$$

$$+ 2(u+v)\left\{-2(u-v)\frac{\partial}{\partial x} + 2(u+v)\frac{\partial}{\partial y}\right\}\left(\frac{\partial f}{\partial y}\right).$$

This reduces to

$$\frac{\partial^2 f}{\partial v^2} = 2\frac{\partial f}{\partial x} + 2\frac{\partial f}{\partial y} + 4x\frac{\partial^2 f}{\partial x^2} - 8\sqrt{(xy)}\frac{\partial^2 f}{\partial x\,\partial y} + 4y\frac{\partial^2 f}{\partial y^2}.$$

It cannot be too strongly emphasised that in examples of this kind operators such as $\partial/\partial u$, $\partial/\partial v$ must not be replaced by their equivalents in terms of x and y before the stage given by equations (5), (6) and (7). Too early use of the equivalent operator is a common source of error leading to the omission of the first order derivatives $\partial f/\partial x$, $\partial f/\partial y$ in the final results.

2:3 The use of equivalent operators

The working involved in problems on second order derivatives can often be considerably shortened by the use of equivalent operators. We consider the following example as an illustration of the method.

EXAMPLE. *If $f(x,y)$ is any function of x and y, where*

$$x = \cosh(u+v), \quad y = \sinh(u-v),$$

show that

$$\frac{\partial^2 f}{\partial u^2} - \frac{\partial^2 f}{\partial v^2} = 4\sqrt{\{(x^2-1)(y^2+1)\}}\frac{\partial^2 f}{\partial x\,\partial y}.$$

[B. Sc.(Gen.)]

Using the Chain Rule

$$\frac{\partial f}{\partial u} = \frac{\partial f}{\partial x}\frac{\partial x}{\partial u} + \frac{\partial f}{\partial y}\frac{\partial y}{\partial u} = \sinh(u+v)\frac{\partial f}{\partial x} + \cosh(u-v)\frac{\partial f}{\partial y}, \quad (1)$$

$$\frac{\partial f}{\partial v} = \frac{\partial f}{\partial x}\frac{\partial x}{\partial v} + \frac{\partial f}{\partial y}\frac{\partial y}{\partial v} = \sinh(u+v)\frac{\partial f}{\partial x} - \cosh(u-v)\frac{\partial f}{\partial y}. \quad (2)$$

We *could* now find $\partial^2 f/\partial u^2$, $\partial^2 f/\partial v^2$ separately by the methods of § 2:2, but this is unnecessarily long. Instead, we note that equations (1) and (2) give

$$\frac{\partial f}{\partial u} + \frac{\partial f}{\partial v} = 2\sinh(u+v)\frac{\partial f}{\partial x} = 2\sqrt{(x^2-1)}\frac{\partial f}{\partial x}, \quad (3)$$

$$\frac{\partial f}{\partial u} - \frac{\partial f}{\partial v} = 2\cosh(u-v)\frac{\partial f}{\partial y} = 2\sqrt{(1+y^2)}\frac{\partial f}{\partial y}. \quad (4)$$

Further, for any function f of u and v, we have

$$\left(\frac{\partial}{\partial u}+\frac{\partial}{\partial v}\right)\left(\frac{\partial f}{\partial u}-\frac{\partial f}{\partial v}\right)=\frac{\partial^2 f}{\partial u^2}-\frac{\partial^2 f}{\partial u\,\partial v}+\frac{\partial^2 f}{\partial v\,\partial u}-\frac{\partial^2 f}{\partial v^2}$$
$$=\frac{\partial^2 f}{\partial u^2}-\frac{\partial^2 f}{\partial v^2}.$$

From equations (3),

$$\frac{\partial}{\partial u}+\frac{\partial}{\partial v}=2\sqrt{(x^2-1)}\frac{\partial}{\partial x}. \tag{5}$$

Hence, using equations (5) and (4),

$$\left(\frac{\partial}{\partial u}+\frac{\partial}{\partial v}\right)\left(\frac{\partial f}{\partial u}-\frac{\partial f}{\partial v}\right)=2\sqrt{(x^2-1)}\frac{\partial}{\partial x}\left\{2\sqrt{(1+y^2)}\frac{\partial f}{\partial y}\right\}.$$

Thus, $$\frac{\partial^2 f}{\partial u^2}-\frac{\partial^2 f}{\partial v^2}=4\sqrt{\{(x^2-1)(1+y^2)\}}\frac{\partial^2 f}{\partial x\,\partial y}.$$

The reader should now work out $\partial^2 f/\partial u^2$, $\partial^2 f/\partial v^2$ separately and compare the length of the two methods.

Exercises 2:3

1. If $x = \cosh u \cos v$, $y = \sinh u \sin v$, and V is any function of x and y, show that

$$(\sinh^2 u\cos^2 v+\cosh^2 u\sin^2 v)\left(\frac{\partial^2 V}{\partial x^2}+\frac{\partial^2 V}{\partial y^2}\right)=\frac{\partial^2 V}{\partial u^2}+\frac{\partial^2 V}{\partial v^2}.$$

[B.Sc.(Eng.)]

2. If $x = e^v \sec u$, $y = e^v \tan u$, and V is any function of x and y, show that

$$\cos u\left(\frac{\partial^2 V}{\partial u\,\partial v}-\frac{\partial V}{\partial u}\right)=xy\left(\frac{\partial^2 V}{\partial x^2}+\frac{\partial^2 V}{\partial y^2}\right)+(x^2+y^2)\frac{\partial^2 V}{\partial x\,\partial y}.$$

3. If the equation

$$\frac{\partial^2 V}{\partial x^2}+\frac{\partial^2 V}{\partial y^2}=0$$

is satisfied by $V = f(y/x)$, show that, given $u = y/x$,

$$(1+u^2)\,\frac{d^2V}{du^2}+2u\frac{dV}{du}=0.$$

Hence find V in terms of x and y.

4. If $u = yz$, $v = zx$, $w = xy$, and if f is a function of x, y, z, and hence of u, v, w, show that

(i) $x\dfrac{\partial f}{\partial x} = v\dfrac{\partial f}{\partial v} + w\dfrac{\partial f}{\partial w}$,

(ii) $x^2\dfrac{\partial^2 f}{\partial x^2} + y^2\dfrac{\partial^2 f}{\partial y^2} + z^2\dfrac{\partial^2 f}{\partial z^2}$

$$= 2\left[u^2\frac{\partial^2 f}{\partial u^2} + v^2\frac{\partial^2 f}{\partial v^2} + w^2\frac{\partial^2 f}{\partial w^2} + uv\frac{\partial^2 f}{\partial u\,\partial v} + vw\frac{\partial^2 f}{\partial v\,\partial w} + wu\frac{\partial^2 f}{\partial w\,\partial u}\right].$$

[B.Sc.(Eng.)]

2:4 Two important transformations

The *Laplacian operator* is important in many branches of applied mathematics and physics. In two variables, it is

$$\frac{\partial^2}{\partial x^2} + \frac{\partial^2}{\partial y^2},$$

and in three variables

$$\frac{\partial^2}{\partial x^2} + \frac{\partial^2}{\partial y^2} + \frac{\partial^2}{\partial z^2}.$$

Apart from Laplace's equations themselves, namely

$$\frac{\partial^2 f}{\partial x^2} + \frac{\partial^2 f}{\partial y^2} = 0, \quad \frac{\partial^2 f}{\partial x^2} + \frac{\partial^2 f}{\partial y^2} + \frac{\partial^2 f}{\partial z^2} = 0,$$

we have the wave equation

$$\frac{\partial^2 f}{\partial x^2} + \frac{\partial^2 f}{\partial y^2} + \frac{\partial^2 f}{\partial z^2} = \frac{1}{c^2}\frac{\partial^2 f}{\partial t^2}$$

and the corresponding form in two variables x and y. In order to obtain solutions of these and other equations involving the Laplacian operator which will satisfy the conditions of a given problem, it is frequently necessary to change to new variables. We consider now two very useful transformations of this operator.

(i) *The plane polar transformation*

The plane polar transformation $x = \rho\cos\psi$, $y = \rho\sin\psi$ applied to the expression $\partial^2 f/\partial x^2 + \partial^2 f/\partial y^2$ gives

$$\frac{\partial^2 f}{\partial\rho^2} + \frac{1}{\rho^2}\frac{\partial^2 f}{\partial\psi^2} + \frac{1}{\rho}\frac{\partial f}{\partial\rho}.$$

The proof of this is a valuable exercise in partial differentiation.

By the Chain Rule, we have

$$\frac{\partial f}{\partial x} = \frac{\partial f}{\partial\rho}\frac{\partial\rho}{\partial x} + \frac{\partial f}{\partial\psi}\frac{\partial\psi}{\partial x}, \quad \frac{\partial f}{\partial y} = \frac{\partial f}{\partial\rho}\frac{\partial\rho}{\partial y} + \frac{\partial f}{\partial\psi}\frac{\partial\psi}{\partial y}. \tag{2}$$

Since $\rho^2 = x^2 + y^2$ and $\tan\psi = y/x$, we obtain, on differentiating partially with respect to x and y,

$$2\rho\frac{\partial\rho}{\partial x} = 2x, \quad 2\rho\frac{\partial\rho}{\partial y} = 2y, \quad \sec^2\psi\frac{\partial\psi}{\partial x} = -\frac{y}{x^2}, \quad \sec^2\psi\frac{\partial\psi}{\partial y} = \frac{1}{x}.$$

These equations give

$$\frac{\partial\rho}{\partial x} = \frac{x}{\rho} = \cos\psi, \quad \frac{\partial\rho}{\partial y} = \frac{y}{\rho} = \sin\psi, \tag{3}$$

$$\frac{\partial\psi}{\partial x} = -\frac{y}{x^2\sec^2\psi} = -\frac{1}{\rho}\sin\psi, \quad \frac{\partial\psi}{\partial y} = \frac{1}{x}\sec^2\psi = \frac{1}{\rho}\cos\psi. \tag{4}$$

So, from (2),

$$\frac{\partial f}{\partial x} = \cos\psi\frac{\partial f}{\partial\rho} - \frac{1}{\rho}\sin\psi\frac{\partial f}{\partial\psi}, \tag{5}$$

$$\frac{\partial f}{\partial y} = \sin\psi\frac{\partial f}{\partial\rho} + \frac{1}{\rho}\cos\psi\frac{\partial f}{\partial\psi}. \tag{6}$$

So,
$$\frac{\partial^2 f}{\partial x^2} = \frac{\partial}{\partial x}\left(\frac{\partial f}{\partial x}\right) = \frac{\partial}{\partial x}\left(\cos\psi\frac{\partial f}{\partial\rho} - \frac{1}{\rho}\sin\psi\frac{\partial f}{\partial\psi}\right). \tag{7}$$

In (7), no further differentiation with respect to x is possible on the right; so we replace $\partial/\partial x$ by its equivalent operator from (5),

$$\cos\psi\frac{\partial}{\partial\rho} - \frac{1}{\rho}\sin\psi\frac{\partial}{\partial\psi}.$$

Thus, $$\frac{\partial^2 f}{\partial x^2} = \left(\cos\psi\frac{\partial}{\partial\rho} - \frac{1}{\rho}\sin\psi\frac{\partial}{\partial\psi}\right)\left(\cos\psi\frac{\partial f}{\partial\rho} - \frac{1}{\rho}\sin\psi\frac{\partial f}{\partial\psi}\right).$$

Care must now be taken to include all possible terms in the expansion of the right-hand side.

$$\frac{\partial^2 f}{\partial x^2} = \cos\psi\left(\cos\psi\frac{\partial^2 f}{\partial\rho^2} + \frac{1}{\rho^2}\sin\psi\frac{\partial f}{\partial\psi} - \frac{1}{\rho}\sin\psi\frac{\partial^2 f}{\partial\rho\,\partial\psi}\right) -$$
$$-\frac{1}{\rho}\sin\psi\left(-\sin\psi\frac{\partial f}{\partial\rho} + \cos\psi\frac{\partial^2 f}{\partial\rho\,\partial\psi} -\right.$$
$$\left.-\frac{1}{\rho}\cos\psi\frac{\partial f}{\partial\psi} - \frac{1}{\rho}\sin\psi\frac{\partial^2 f}{\partial\psi^2}\right).$$

So, $$\frac{\partial^2 f}{\partial x^2} = \cos^2\psi\frac{\partial^2 f}{\partial\rho^2} - \frac{2}{\rho}\cos\psi\sin\psi\frac{\partial^2 f}{\partial\rho\,\partial\psi} + \frac{\sin^2\psi}{\rho^2}\frac{\partial^2 f}{\partial\psi^2} +$$
$$+\frac{\sin^2\psi}{\rho}\frac{\partial f}{\partial\rho} + \frac{2}{\rho^2}\sin\psi\cos\psi\frac{\partial f}{\partial\psi}.$$

In an exactly similar way, we find

$$\frac{\partial^2 f}{\partial y^2} = \sin^2\psi\frac{\partial^2 f}{\partial\rho^2} + \frac{2}{\rho}\cos\psi\sin\psi\frac{\partial^2 f}{\partial\rho\,\partial\psi} + \frac{\cos^2\psi}{\rho^2}\frac{\partial^2 f}{\partial x^2} +$$
$$+\frac{\cos^2\psi}{\rho}\frac{\partial f}{\partial\rho} - \frac{2}{\rho^2}\sin\psi\cos\psi\frac{\partial f}{\partial\psi}.$$

Hence,
$$\frac{\partial^2 f}{\partial x^2} + \frac{\partial^2 f}{\partial y^2} = \frac{\partial^2 f}{\partial\rho^2} + \frac{1}{\rho^2}\frac{\partial^2 f}{\partial\psi^2} + \frac{1}{\rho}\frac{\partial f}{\partial\rho}. \tag{8}$$

(ii) *The spherical polar transformation*

The spherical polar transformation

$$x = r\sin\theta\cos\psi, \quad y = r\sin\theta\sin\psi, \quad z = r\cos\theta$$

applied to the expression

$$\frac{\partial^2 f}{\partial x^2}+\frac{\partial^2 f}{\partial y^2}+\frac{\partial^2 f}{\partial z^2}$$

produces

$$\frac{\partial^2 f}{\partial r^2}+\frac{1}{r^2}\frac{\partial^2 f}{\partial \theta^2}+\frac{1}{r^2\sin^2\theta}\frac{\partial^2 f}{\partial \psi^2}+\frac{2}{r}\frac{\partial f}{\partial r}+\frac{\cot\theta}{r^2}\frac{\partial f}{\partial r}.$$

To find $\partial^2 f/\partial x^2$, $\partial^2 f/\partial y^2$, $\partial^2 f/\partial z^2$ all separately is a very lengthy process. Instead, we make use of the plane polar transformation. Put $u = r\sin\theta$; then $x = u\cos\psi$, $y = u\sin\psi$. So using equation (8), with ρ replaced by u,

$$\frac{\partial^2 f}{\partial x^2}+\frac{\partial^2 f}{\partial y^2}=\frac{\partial^2 f}{\partial u^2}+\frac{1}{u^2}\frac{\partial^2 f}{\partial \psi^2}+\frac{1}{u}\frac{\partial f}{\partial u}. \tag{9}$$

Also $z = r\cos\theta$, $u = r\sin\theta$; so using (8) again,

$$\frac{\partial^2 f}{\partial z^2}+\frac{\partial^2 f}{\partial u^2}=\frac{\partial^2 f}{\partial r^2}+\frac{1}{r^2}\frac{\partial^2 f}{\partial \theta^2}+\frac{1}{r}\frac{\partial f}{\partial r}. \tag{10}$$

Adding (9) and (10), and noting that $\partial^2 f/\partial u^2$ occurs on both sides,

$$\frac{\partial^2 f}{\partial x^2}+\frac{\partial^2 f}{\partial y^2}+\frac{\partial^2 f}{\partial z^2}=\frac{\partial^2 f}{\partial r^2}+\frac{1}{r^2}\frac{\partial^2 f}{\partial \theta^2}+\frac{1}{u^2}\frac{\partial^2 f}{\partial \psi^2}+\frac{1}{r}\frac{\partial f}{dr}+\frac{1}{u}\frac{\partial f}{\partial u}. \tag{11}$$

In equation (11) the term

$$\frac{1}{u}\frac{\partial f}{\partial u}$$

on the right needs further attention. Comparing the transformation $z = r\cos\theta$, $u = r\sin\theta$ with the plane polar one in which x, y replace z, u respectively, and using equation (6), we find

$$\frac{\partial f}{\partial u}=\sin\theta\frac{\partial f}{\partial \theta}+\frac{1}{r}\cos\theta\frac{\partial f}{\partial \theta}. \tag{12}$$

Hence, from (11) and (12),

$$\frac{\partial^2 f}{\partial x^2}+\frac{\partial^2 f}{\partial y^2}+\frac{\partial^2 f}{\partial z^2}$$
$$=\frac{\partial^2 f}{\partial r^2}+\frac{1}{r^2}\frac{\partial^2 f}{\partial \theta^2}+\frac{1}{r^2\sin^2\theta}\frac{\partial^2 f}{\partial \psi^2}+\frac{2}{r}\frac{\partial f}{\partial r}+\frac{\cot\theta}{r^2}\frac{\partial f}{\partial \theta}. \quad (13)$$

Exercises 2:4

1. If $x = \rho\cos\psi$, $y = \rho\sin\psi$ and V is a function of x and y, show that

(i) $$x^2\frac{\partial V}{\partial x}+y^2\frac{\partial V}{\partial y}$$
$$=(\cos^3\psi+\sin^3\psi)\rho^2\frac{\partial V}{\partial \rho}+\rho\cos\psi\sin\psi(\sin\psi-\cos\psi)\frac{\partial V}{\partial \psi},$$

(ii) $$\frac{\partial^2 V}{\partial x^2}\frac{\partial^2 V}{\partial y^2}-\left(\frac{\partial^2 V}{\partial x\,\partial y}\right)^2$$
$$=\frac{1}{\rho^2}\frac{\partial^2 V}{\partial \rho^2}\left(\frac{\partial^2 V}{\partial \psi^2}+\rho\frac{\partial V}{\partial \rho}\right)-\frac{1}{\rho^2}\left(\frac{\partial^2 V}{\partial \rho\,\partial \psi}-\frac{1}{\rho}\frac{\partial V}{\partial \psi}\right)^2.$$

(In (ii) work from the right-hand side by finding $\partial V/\partial \rho$, $\partial V/\partial \psi$, etc., in terms of $\partial V/\partial x$, $\partial V/\partial y$, etc.) [B.Sc.(Eng.)]

2. If ϕ is a function of ρ and ψ such that $\phi = f(\rho)\sin\psi$, where $f(\rho)$ is an arbitrary function of ρ, and if $x = \rho\cos\psi$, $y = \rho\sin\psi$, show that

$$\left(x^2+y^2\right)\left(\frac{\partial^2\phi}{\partial x^2}+\frac{\partial^2\phi}{\partial y^2}\right)=\left(\rho^2\frac{d^2 f}{d\rho^2}+\rho\frac{df}{d\rho}-f\right)\sin\psi.$$

Find the general solution of the differential equation

$$\rho^2\frac{d^2 f}{d\rho^2}+\rho\frac{df}{d\rho}-f=0,$$

and hence write down the general expression for ϕ when $\phi = f(\rho)\sin\psi$ is a solution of Laplace's equation.

3. If $2x = u^2 - v^2$, $y = uv$, $z = w$, show that Laplace's equation

$$\frac{\partial^2\phi}{\partial x^2}+\frac{\partial^2\phi}{\partial y^2}+\frac{\partial^2\phi}{\partial z^2}=0$$

transforms into

$$\frac{\partial^2\phi}{\partial u^2}+\frac{\partial^2\phi}{\partial v^2}+(u^2+v^2)\frac{\partial^2\phi}{\partial w^2}=0.$$

[u, v, w are *parabolic cylindrical coordinates;* see Exercises 7:8, No. 4.]

2:5 Total differentials and their use

For a function $y = f(x)$ of one variable x, the differential $\mathrm{d}y$ is defined by the equation

$$\mathrm{d}y = f'(x)\,\mathrm{d}x, \tag{1}$$

where $\mathrm{d}x$ is an arbitrary increment, and $f'(x) = \mathrm{d}f/\mathrm{d}x$.

For a function $z = f(x, y)$ of two variables x and y, the total or exact differential $\mathrm{d}z$ is defined by the equation

$$\mathrm{d}z = \frac{\partial z}{\partial x}\,\mathrm{d}x + \frac{\partial z}{\partial y}\,\mathrm{d}y, \tag{2}$$

where $\mathrm{d}x$, $\mathrm{d}y$ are arbitrary increments in x and y.

For a function $f(x,y,z)$ of three variables the total differential $\mathrm{d}f$ is given by

$$\mathrm{d}f = \frac{\partial f}{\partial x}\mathrm{d}x + \frac{\partial f}{\partial y}\mathrm{d}y + \frac{\partial f}{\partial z}\mathrm{d}z, \tag{3}$$

and so on for any number of variables.

We make use of the total differential when working with implicit functions. Suppose, for example, that u and v are given as functions of x and y by the relations

$$x = u + v\log u, \quad y = v - u\log v. \tag{4}$$

Then u and v are given *implicitly* in terms of x and y, since clearly we cannot solve the equations to obtain u, v explicitly expressed in terms of x and y. The relations (4) can be thought of in two ways; first as giving x, y explicitly as functions of u and v and secondly as giving u, v implicitly as functions of x and y. Thinking of u, v as functions of x and y, we must be able to find the partial derivatives $\partial u/\partial x$, $\partial u/\partial y$, $\partial v/\partial x$, $\partial v/\partial y$. We begin by writing down expressions for $\mathrm{d}x$ and $\mathrm{d}y$ using the definition (2). Then

$$\mathrm{d}x = \frac{\partial x}{\partial u}\mathrm{d}u + \frac{\partial x}{\partial v}\mathrm{d}v = \left(1 + \frac{v}{u}\right)\mathrm{d}u + \log u\,\mathrm{d}v, \tag{5}$$

$$\mathrm{d}y = \frac{\partial y}{\partial u}\mathrm{d}u + \frac{\partial y}{\partial v}\mathrm{d}v = -\log v\,\mathrm{d}u + \left(1 - \frac{u}{v}\right)\mathrm{d}v. \tag{6}$$

Now equations (5) and (6) may be thought of as two equations for du and dv. Suppose we solve for du by eliminating dv. Then

$$(v^2 - u^2 + uv \log u \log v)\,\mathrm{d}u = u(v-u)\,\mathrm{d}x - uv \log u\,\mathrm{d}y. \quad (7)$$

But u is a function of x and y; so using the definition (2) again, we must have

$$\mathrm{d}u = \frac{\partial u}{\partial x}\mathrm{d}x + \frac{\partial u}{\partial y}\mathrm{d}y. \quad (8)$$

Comparing equations (7) and (8), we find

$$\frac{\partial u}{\partial x} = \frac{u(v-u)}{v^2 - u^2 + uv \log u \log v}, \quad \frac{\partial u}{\partial y} = \frac{-uv \log u}{v^2 - u^2 + uv \log u \log v}.$$

Now we solve equations (5) and (6) for dv by eliminating du. Then

$$(v^2 - u^2 + uv \log u \log v)\,\mathrm{d}v = uv \log v\,\mathrm{d}x + v(u+v)\,\mathrm{d}y. \quad (9)$$

But v is a function of x and y; so

$$\mathrm{d}v = \frac{\partial v}{\partial x}\mathrm{d}x + \frac{\partial v}{\partial y}\mathrm{d}y. \quad (10)$$

Comparing equations (9) and (10)

$$\frac{\partial v}{\partial x} = \frac{uv \log v}{v^2 - u^2 + uv \log u \log v}, \quad \frac{\partial v}{\partial y} = \frac{v(u+v)}{v^2 - u^2 + uv \log u \log v}.$$

The method just described is particularly useful when all four first derivatives are required. If we want to find only $\partial u/\partial x$ and $\partial v/\partial x$, a shorter method is available. Starting with equations (4)

$$x = u + v \log u, \quad y = v - u \log v,$$

we differentiate partially with respect to x, keeping y constant. Then

$$1 = \frac{\partial u}{\partial x} + \frac{v}{u}\frac{\partial u}{\partial x} + \log u \frac{\partial v}{\partial x}, \quad (11)$$

$$0 = \frac{\partial v}{\partial x} - \frac{u}{v}\frac{\partial v}{\partial x} - \log v \frac{\partial u}{\partial x}. \quad (12)$$

Equations (11), (12) may now be solved for $\partial u/\partial x$, $\partial v/\partial x$. Similarly, if we want only $\partial u/\partial y$, $\partial v/\partial y$, then partial differentiation of (4) with respect to y will give two equations which can be solved for $\partial u/\partial y$, $\partial v/\partial y$.

Exercises 2:5

1. Given the functions

$$f_1(x, y, u, v) = u^3 + v^3 + x^3 - 3y,$$

$$f_2(x, y, u, v) = u^2 + v^2 + y^2 + 2x,$$

write down expressions for the total differentials df_1, df_2. Hence, if the equations $f_1 = 0$, $f_2 = 0$ define u and v as functions of x and y, find $\partial u/\partial x$, $\partial u/\partial y$, $\partial v/\partial x$, $\partial v/\partial y$. (If $f_1 = 0$, then $df_1 = 0$; similarly, $df_2 = 0$. This gives two equations to solve for du, dv.)

2. If $f(y - z)$ is an arbitrary function of $(y - z)$, and if $z^2 = x^2 + f(y - z)$, show that

$$z\frac{\partial z}{\partial x} + x\frac{\partial z}{\partial y} = x.$$

3. If the equations $u^6 + v^9 = x^4 + y^8$, $u^4 + v^6 = x^3 + y^6$ are considered as defining u and v as functions of x and y, show that

$$\frac{\partial u}{\partial x} = \frac{x^2(8x - 9v^3)}{12u^3(u^2 - v^3)}.$$

Find also $\partial x/\partial u$, when the equations are considered as defining x and y in terms of u and v.

2:6 Jacobians

If x and y are functions of u and v, the quantity

$$\frac{\partial x}{\partial u}\frac{\partial y}{\partial v} - \frac{\partial x}{\partial v}\frac{\partial y}{\partial u}$$

is the *Jacobian* of x and y with respect to u and v. Jacobians play an important part in the theory of change of variable and we shall use them later when dealing with multiple integrals. The Jacobian here defined is written briefly as

$$\frac{\partial(x, y)}{\partial(u, v)}.$$

Determinant notation provides a useful form; for

$$\frac{\partial(x, y)}{\partial(u, v)} = \begin{vmatrix} \dfrac{\partial x}{\partial u} & \dfrac{\partial x}{\partial v} \\ \dfrac{\partial y}{\partial u} & \dfrac{\partial y}{\partial v} \end{vmatrix}. \tag{1}$$

For three variables x, y, z which are each functions of u, v, w the corresponding Jacobian is

$$\frac{\partial(x, y, z)}{\partial(u, v, w)} = \begin{vmatrix} \dfrac{\partial x}{\partial u} & \dfrac{\partial x}{\partial v} & \dfrac{\partial x}{\partial w} \\ \dfrac{\partial y}{\partial u} & \dfrac{\partial y}{\partial v} & \dfrac{\partial y}{\partial w} \\ \dfrac{\partial z}{\partial u} & \dfrac{\partial z}{\partial v} & \dfrac{\partial z}{\partial w} \end{vmatrix}. \tag{2}$$

The definition extends to any number of variables.

If the equations giving x and y as functions of u and v are treated as also defining u and v as functions of x and y, then we obtain the Jacobian

$$\frac{\partial(u, v)}{\partial(x, y)} = \begin{vmatrix} \dfrac{\partial u}{\partial x} & \dfrac{\partial u}{\partial y} \\ \dfrac{\partial v}{\partial x} & \dfrac{\partial v}{\partial y} \end{vmatrix}. \tag{3}$$

An interesting and important result is that

$$\frac{\partial(x, y)}{\partial(u, v)} = 1 \bigg/ \frac{\partial(u, v)}{\partial(x, y)}. \tag{4}$$

To prove this, suppose that x and y are given as differentiable functions of u and v by the equations $x = x(u,v)$, $y = y(u,v)$.

Suppose further that these equations also define u, v as differentiable functions of x and y. Write

$$J = \frac{\partial(x, y)}{\partial(u, v)} = \frac{\partial x}{\partial u}\frac{\partial y}{\partial v} - \frac{\partial x}{\partial v}\frac{\partial y}{\partial u}. \tag{5}$$

Since x and y are differentiable functions of u and v,

$$\left.\begin{aligned} dx &= \frac{\partial x}{\partial u}du + \frac{\partial x}{\partial v}dv, \\ dy &= \frac{\partial y}{\partial u}du + \frac{\partial y}{\partial v}dv. \end{aligned}\right\} \tag{6}$$

Solving equations (6) for du and dv we find

$$\left.\begin{aligned} \left(\frac{\partial x}{\partial u}\frac{\partial y}{\partial v} - \frac{\partial x}{\partial v}\frac{\partial y}{\partial u}\right)du &= \frac{\partial y}{\partial v}dx - \frac{\partial x}{\partial v}dy, \\ \left(\frac{\partial x}{\partial u}\frac{\partial y}{\partial v} - \frac{\partial x}{\partial v}\frac{\partial y}{\partial u}\right)dv &= -\frac{\partial y}{\partial u}dx + \frac{\partial x}{\partial u}dy. \end{aligned}\right\} \tag{7}$$

But u and v are differentiable functions of x and y; so

$$\left.\begin{aligned} du &= \frac{\partial u}{\partial x}dx + \frac{\partial u}{\partial y}dy, \\ dv &= \frac{\partial v}{\partial x}dx + \frac{\partial v}{\partial y}dy. \end{aligned}\right\} \tag{8}$$

Comparing equations (7) and (8), we find

$$\left.\begin{aligned} \frac{\partial u}{\partial x} &= \frac{1}{J}\frac{\partial y}{\partial v}, \quad \frac{\partial u}{\partial y} = -\frac{1}{J}\frac{\partial x}{\partial v}, \\ \frac{\partial v}{\partial x} &= -\frac{1}{J}\frac{\partial y}{\partial u}, \quad \frac{\partial v}{\partial y} = \frac{1}{J}\frac{\partial x}{\partial u}. \end{aligned}\right\} \tag{9}$$

Hence, from (9),

$$\frac{\partial(u, v)}{\partial(x, y)} = \frac{\partial u}{\partial x}\frac{\partial v}{\partial y} - \frac{\partial u}{\partial y}\frac{\partial v}{\partial x}$$

$$= \frac{1}{J^2}\left\{\frac{\partial y}{\partial v}\frac{\partial x}{\partial u} - \frac{\partial x}{\partial v}\frac{\partial y}{\partial u}\right\}$$

$$= \frac{1}{J}.$$

So $$J = \frac{\partial(x, y)}{\partial(u, v)} = 1\bigg/\frac{\partial(u, v)}{\partial(x, y)}.$$

The reader should note that this relation between the Jacobians is the result for functions of two variables which corresponds to the well-known result

$$\frac{\mathrm{d}y}{\mathrm{d}x} = 1\bigg/\frac{\mathrm{d}x}{\mathrm{d}y}$$

for functions of one variable. The equations (9) show that in general $\partial u/\partial x$ is *not* equal to $1/(\partial x/\partial u)$. For, from (9), we find

$$\frac{\partial u}{\partial x} = \frac{1}{J}\frac{\partial y}{\partial v}.$$

Similarly results hold for $\partial u/\partial y$, $\partial v/\partial x$, $\partial v/\partial y$.

The inverse relation (4) between Jacobians is of practical use if, for example, $\partial(u, v)/\partial(x, y)$ is required and the derivatives of x, y with respect to u, v are more easily found than those of u, v with respect to x and y. Consider the following case: if $x = u + v$, $y = uv$, find $\partial(u, v)/\partial(x, y)$. Here

$$\frac{\partial(x, y)}{\partial(u, v)} = \frac{\partial x}{\partial u}\frac{\partial y}{\partial v} - \frac{\partial x}{\partial v}\frac{\partial y}{\partial u} = u - v.$$

Hence $$\frac{\partial(u, v)}{\partial(x, y)} = 1\bigg/\frac{\partial(x, y)}{\partial(u, v)} = \frac{1}{(u - v)}.$$

Exercises 2:6

1. If $u = x + y + z$, $v = x^2 + y^2 + z^2$, $w = x^3 + y^3 + z^3 - 3xyz$, show that

$$\frac{\partial(u, v, w)}{\partial(x, y, z)} = 0.$$

2. (i) If $x = \rho \cos \psi$, $y = \rho \sin \psi$, evaluate $\partial(x, y)/\partial(\rho, \psi)$, and verify that

$$\frac{\partial(\rho, \psi)}{\partial(x, y)} = 1\bigg/\frac{\partial(x, y)}{\partial(\rho, \psi)}.$$

(ii) If $x = r \sin \theta \cos \psi$, $y = r \sin \theta \sin \psi$, $z = r \cos \theta$, evaluate

$$\frac{\partial(x, y, z)}{\partial(r, \theta, \psi)}.$$

(iii) If $2x = u^2 - v^2$, $y = uv$, $z = w$, evaluate

$$\frac{\partial(x, y, z)}{\partial(u, v, w)}$$

and verify that

$$\frac{\partial(u, v, w)}{\partial(x, y, z)} = 1\bigg/\frac{\partial(x, y, z)}{\partial(u, v, w)}.$$

3. If x and y are functions of u and v, and u, v are changed to new variables ξ, η, show that

$$\frac{\partial(x, y)}{\partial(\xi, \eta)} = \frac{\partial(x, y)}{\partial(u, v)}\frac{\partial(u, v)}{\partial(\xi, \eta)}.$$

[Use the rule for multiplication of determinants and the Chain Rule.]

Miscellaneous Exercises 2

1. If $V(x, y)$ is a function of x and y defined by the equation

$$V = x\phi(y/x) + \psi(y/x),$$

where ϕ, ψ are arbitrary functions of y/x, show that

$$x^2\frac{\partial^2 V}{\partial x^2} + 2xy\frac{\partial^2 V}{\partial x\,\partial y} + y^2\frac{\partial^2 V}{\partial y^2} = 0.$$

2. If $f(x)$, $g(y)$ are arbitrary twice differentiable functions of x and y respectively, and if $z(x, y)$ is defined by

$$z = f'(x)g'(y)/\{f(x) + g(y)\}^2,$$

where $f'(x) = \mathrm{d}f/\mathrm{d}x$, $g'(y) = \mathrm{d}g/\mathrm{d}y$, show that

$$\frac{\partial^2(\log z)}{\partial x\,\partial y} = 2z.$$

3. If $u = 1/x + 1/y$, $v = 1/(x+y)$, and z is a function of x and y, prove that

$$\frac{\partial z}{\partial u} = \left(\frac{x^2y^2}{x^2-y^2}\right)\left(\frac{\partial z}{\partial x} - \frac{\partial z}{\partial y}\right), \quad \frac{\partial z}{\partial v} = \left(\frac{x+y}{x-y}\right)\left(y^2\frac{\partial z}{\partial y} - x^2\frac{\partial z}{\partial x}\right).$$

Hence find solutions of the equations

$$\text{(i)}\ \frac{\partial z}{\partial x} - \frac{\partial z}{\partial y} = x^2 - y^2, \quad \text{(ii)}\ y^2\frac{\partial z}{\partial y} - x^2\frac{\partial z}{\partial x} = xy(x-y).$$

[(i) transforms to $\partial z/\partial u = 1/u^2v^2$; solve this equation for z. Similarly for (ii).]

4. Given that ϕ is a function of x and y, where $x^2 + y^2 = u$, $2xy = v$, show that the equation

$$y\frac{\partial \phi}{\partial x} - x\frac{\partial \phi}{\partial y} = \frac{(y^2-x^2)^3}{xy}$$

transforms into

$$\frac{\partial \phi}{\partial v} = \frac{u^2-v^2}{v}.$$

Hence find ϕ in terms of x and y.

5. If V is a function of x and y, where $u(x^2+y^2) = x$, $v(x^2+y^2) = y$, show that

$$\frac{\partial^2 V}{\partial x^2} + \frac{\partial^2 V}{\partial y^2} = (u^2+v^2)^2\left(\frac{\partial^2 V}{\partial u^2} + \frac{\partial^2 V}{\partial v^2}\right).$$

6. If $x + y = (u+v)^n$, $x - y = (u-v)^n$, show that

$$(u^2-v^2)\left(\frac{\partial^2 f}{\partial u^2} - \frac{\partial^2 f}{\partial v^2}\right) = n^2(x^2-y^2)\left(\frac{\partial^2 f}{\partial x^2} - \frac{\partial^2 f}{\partial y^2}\right),$$

where $f(x, y)$ is any function of x and y.

7. If y is defined as a function of x by the equation $f(x, y) = 0$, show that

$$\text{(i)}\ \left(\frac{\partial f}{\partial y}\right)^3\frac{\mathrm{d}^2 y}{\mathrm{d}x^2} = -\frac{\partial^2 f}{\partial x^2}\left(\frac{\partial f}{\partial y}\right)^2 + 2\frac{\partial^2 f}{\partial x\,\partial y}\frac{\partial f}{\partial x}\frac{\partial f}{\partial y} - \frac{\partial^2 f}{\partial y^2}\left(\frac{\partial f}{\partial x}\right)^2,$$

$$\text{(ii)}\ \left(\frac{\partial f}{\partial y}\right)^3\frac{\mathrm{d}^2 y}{\mathrm{d}x^2} = \left(\frac{\partial f}{\partial x}\right)^3\frac{\mathrm{d}^2 x}{\mathrm{d}y^2}.$$

[B.Sc.(Eng.)]

8. (i) The function $H(x, y, z)$ is a homogeneous function of degree n in x, y, z. Prove that

$$x^2\frac{\partial^2 H}{\partial x^2} + y^2\frac{\partial^2 H}{\partial y^2} + z^2\frac{\partial^2 H}{\partial z^2} + 2xy\frac{\partial^2 H}{\partial x\,\partial y} + + 2yz\frac{\partial^2 H}{\partial y\,\partial z} + 2zx\frac{\partial^2 H}{\partial z\,\partial x} = n(n-1)H.$$

If, in addition, $H(x, y, z)$ satisfies Laplace's equation and if $r^2 = x^2 + y^2 + z^2$, show that

$$\left(\frac{\partial^2}{\partial x^2} + \frac{\partial^2}{\partial y^2} + \frac{\partial^2}{\partial z^2}\right)(Hr^{-2n-1}) = 0.$$

(ii) Given that $\theta = \tan^{-1}u$, where u is a homogeneous function of degree 2 in x, y, z, prove that

$$x\frac{\partial\theta}{\partial x} + y\frac{\partial\theta}{\partial y} + z\frac{\partial\theta}{\partial z} = \sin 2\theta.$$ [B.Sc.(Eng.)]

9. If u and v are functions of x and y satisfying the equations

$$\frac{\partial u}{\partial x} = \frac{\partial v}{\partial y}, \quad \frac{\partial u}{\partial y} = -\frac{\partial v}{\partial x},$$

show that $$\frac{\partial^2 u}{\partial x^2} + \frac{\partial^2 u}{\partial y^2} = 0, \quad \frac{\partial^2 v}{\partial x^2} + \frac{\partial^2 v}{\partial y^2} = 0.$$

If θ is any function of u and v, deduce that

$$\frac{\partial^2\theta}{\partial x^2} + \frac{\partial^2\theta}{\partial y^2} = \left\{\left(\frac{\partial u}{\partial x}\right)^2 + \left(\frac{\partial u}{\partial y}\right)^2\right\}\left\{\frac{\partial^2\theta}{\partial u^2} + \frac{\partial^2\theta}{\partial v^2}\right\}.$$ [B.Sc.(Eng.)]

10. If $u = x^3 - 3xy^2$, $v = 3x^2y - y^3$, and V is an arbitrary function of x and y, and hence of u and v, show that

$$(x^2 + y^2)\left(\frac{\partial^2 V}{\partial x^2} + \frac{\partial^2 V}{\partial y^2}\right) = 9(u^2+v^2)\left(\frac{\partial^2 V}{\partial u^2} + \frac{\partial^2 V}{\partial v^2}\right).$$

11. The function $F(r, \theta)$ of two independent variables r, θ is transformed to a function $G(u, s)$ of two new variables u, s by the change of variables $r\cos\theta = 1/u$, $\tan\theta = s$. Show that

$$\text{(i)}\ r\frac{\partial F}{\partial r} = -u\frac{\partial G}{\partial u}, \quad \text{(ii)}\ \frac{\partial F}{\partial\theta} = us\frac{\partial G}{\partial u} + (1 + s^2)\frac{\partial G}{\partial s}.$$

If F satisfies the differential equation

$$\cos\theta\frac{\partial^2 F}{\partial r\,\partial\theta} + r\sin\theta\frac{\partial^2 F}{\partial r^2} = 0,$$

prove that

$$\frac{\partial G}{\partial u} = (1 + s^2)^{1/2}\phi(u),$$

where $\phi(u)$ is an arbitrary function of u.

(Show that the given differential equation transforms to

$$s\frac{\partial G}{\partial u} - (1 + s^2)\frac{\partial^2 G}{\partial u\, \partial s} = 0;$$

writing $\partial G/\partial u = H$, say, we now obtain

$$\frac{1}{H}\frac{\partial H}{\partial s} = \frac{s}{1 + s^2},$$

from which the required result follows by integration.) [B.Sc.(Eng.)]

12. If z is a function of x and y and $x = u + v$, $y = uv$, prove that

(i) $(u - v)\dfrac{\partial z}{\partial x} = u\dfrac{\partial z}{\partial u} - v\dfrac{\partial z}{\partial v}$; $(u - v)\dfrac{\partial z}{\partial y} = \dfrac{\partial z}{\partial v} - \dfrac{\partial z}{\partial u}$,

(ii) $\partial(u, v)/\partial(x, y) = 1/(u - v)$,

(iii) $\dfrac{\partial^2 z}{\partial x^2} - y\dfrac{\partial^2 z}{\partial y^2} = \dfrac{u}{u - v}\dfrac{\partial^2 z}{\partial u^2} - \dfrac{v}{u - v}\dfrac{\partial^2 z}{\partial v^2}$. [B.Sc.(Eng.)]

13. The independent variables x, y are transformed into new variables X, Y given by the equations $X = xy$, $Y = 1/y$. If the function $f(x, y)$ is thus transformed into $F(X, Y)$, prove that

(i) $y\dfrac{\partial f}{\partial y}\left(x\dfrac{\partial f}{\partial x} - y\dfrac{\partial f}{\partial y}\right) = Y\dfrac{\partial F}{\partial Y}\left(X\dfrac{\partial F}{\partial X} - Y\dfrac{\partial F}{\partial Y}\right)$,

(ii) $y\left(x\dfrac{\partial^2 f}{\partial x\, \partial y} - y\dfrac{\partial^2 f}{\partial y^2} - \dfrac{\partial f}{\partial y}\right) = Y\left(X\dfrac{\partial^2 F}{\partial X\, \partial Y} - Y\dfrac{\partial^2 F}{\partial Y^2} - \dfrac{\partial F}{\partial Y}\right)$.

[B.Sc.(Eng.)]

14. If f is a function of the independent variables x and y which are changed to independent variables u and v, where $u = e^{y/x}$, $v = x^2 + y^2$, show that

(i) $2uv\, dx = xu\, dv - 2yx^2\, du$; $2uv\, dy = 2x^3\, du + uy\, dv$,

(ii) $x\dfrac{\partial f}{\partial x} + y\dfrac{\partial f}{\partial y} = 2v\dfrac{\partial f}{\partial v}$; $x^3\dfrac{\partial f}{\partial y} - x^2 y\dfrac{\partial f}{\partial x} = uv\dfrac{\partial f}{\partial u}$,

(iii) $2uv^2\dfrac{\partial^2 f}{\partial u\, \partial v} - uv\dfrac{\partial f}{\partial u} = x^3 y\left(\dfrac{\partial^2 f}{\partial y^2} - \dfrac{\partial^2 f}{\partial x^2}\right) + x^2(x^2 - y^2)\dfrac{\partial^2 f}{\partial x\, \partial y}$.

[B.Sc.(Eng.)]

15. For the transformations $x = \rho \cos w$, $y = \rho \sin w$, and $\rho = a \cosh u \cos v$, $z = a \sinh u \sin v$, applied to the function $F(x, y, z) = \phi(\rho, w, z) = \psi(u, v, w)$,

(i) express
$$\frac{\partial^2 F}{\partial x^2} + \frac{\partial^2 F}{\partial y^2}$$

in terms of $\partial^2\phi/\partial\rho^2$, $\partial^2\phi/\partial w^2$, and $\partial\phi/\partial\rho$,

(ii) show that the equation
$$\frac{\partial^2 F}{\partial x^2} + \frac{\partial^2 F}{\partial y^2} + \frac{\partial^2 F}{\partial z^2} = 0$$

transforms to
$$\frac{\partial^2\psi}{\partial u^2} + \frac{\partial^2\psi}{\partial v^2} + \tanh u \frac{\partial\psi}{\partial u} - \tan v \frac{\partial\psi}{\partial v} + \frac{\sinh^2 u + \sin^2 v}{\cosh^2 u \cos^2 v} \frac{\partial^2\psi}{\partial w^2} = 0.$$

Hence find a solution of this equation which is a function of u only. (In (ii), begin by expressing
$$\frac{\partial^2\psi}{\partial u^2} + \frac{\partial^2\psi}{\partial v^2}$$

in terms of $\partial^2\phi/\partial\rho^2$, $\partial^2\phi/\partial z^2$, and then use the result of (i).) [B.Sc.(Eng.)]

16. The variables x, y, z are changed to new variables u, v, w by means of the transformation equations $x = au + bv + cw$, $y = a'u + b'v + c'w$, $z = a''u + b''v + c''w$, where $a, b, c, a', b', c', a'', b'', c''$ are constants related by the equations $a^2 + b^2 + c^2 = 1$, $a'^2 + b'^2 + c'^2 = 1$, $a''^2 + b''^2 + c''^2 = 1$, $aa' + bb' + cc' = 0$, $aa'' + bb'' + cc'' = 0$, $a'a'' + b'b'' + c'c'' = 0$.
If V is any function of x, y, and z, show that

(i) $$\left(\frac{\partial V}{\partial x}\right)^2 + \left(\frac{\partial V}{\partial y}\right)^2 + \left(\frac{\partial V}{\partial z}\right)^2 = \left(\frac{\partial V}{\partial u}\right)^2 + \left(\frac{\partial V}{\partial v}\right)^2 + \left(\frac{\partial V}{\partial w}\right)^2,$$

(ii) $$\frac{\partial^2 V}{\partial x^2} + \frac{\partial^2 V}{\partial y^2} + \frac{\partial^2 V}{\partial z^2} = \frac{\partial^2 V}{\partial u^2} + \frac{\partial^2 V}{\partial v^2} + \frac{\partial^2 V}{\partial w^2}.$$

(The transformation corresponds to a rotation of the axes $Oxyz$ about the origin O, the expressions occurring in (i) and (ii) being *invariant* under this transformation.)

17. (i) If $u = x^n f(Y, Z)$, where $Y = y/x$, $Z = z/x$, show that
$$x\frac{\partial u}{\partial x} + y\frac{\partial u}{\partial y} + z\frac{\partial u}{\partial z} = nu.$$

(ii) If $v = \log(x + r)$, where $r^2 = x^2 + y^2 + z^2$, show that

$$x\frac{\partial v}{\partial x} + y\frac{\partial v}{\partial y} + z\frac{\partial v}{\partial z} = 1.$$

18. If $x = uv$, $y = u/v$, and F is a function of x and y, prove that

$$4\frac{\partial^2 F}{\partial x\,\partial y} = \frac{\partial^2 F}{\partial u^2} + \frac{1}{u}\frac{\partial F}{\partial u} - \frac{v}{u^2}\frac{\partial F}{\partial v} - \frac{v^2}{u^2}\frac{\partial^2 F}{\partial v^2}.$$

Hence find the general solution of the equation $\partial^2 F/\partial x\,\partial y = 0$, when F is a function of $\sqrt{(xy)}$ only.

CHAPTER 3

LINE INTEGRALS

3:1 Line integrals in two dimensions

Most curves in the x–y plane can be given by an equation of the form $f(x, y) = 0$: in practice, parametric forms are often more convenient, with x and y expressed in terms of a variable t, say. For example, the circle $x^2 + y^2 = a^2$ has the parametric equations $x = a \cos t$, $y = a \sin t$; the parabola $y^2 = 4ax$ has parametric equations $x = at^2$, $y = 2at$. We shall write the parametric equations of the curve $f(x, y) = 0$ as $x = \psi(t)$, $y = \phi(t)$.

Consider the integrals $\int g(x, y)\,dx$, $\int g(x, y)\,dy$, where $g(x, y)$ is a given function of x and y, but x, y are restricted so that the point (x, y) lies on the curve $x = \psi(t)$, $y = \phi(t)$.

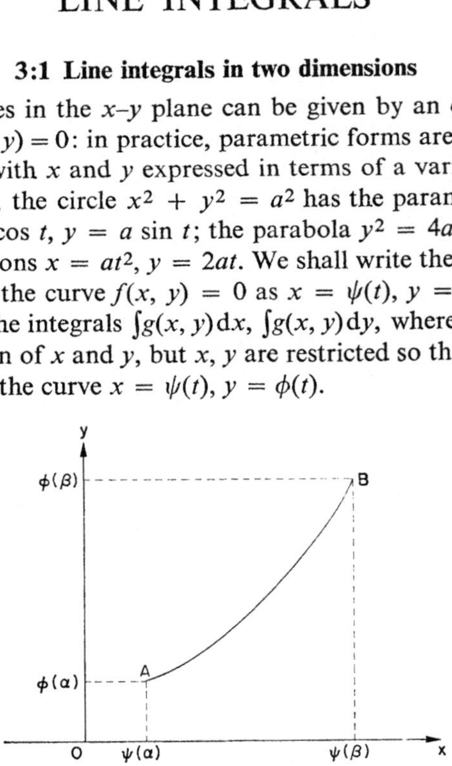

FIG. 1

Suppose the arc AB of the curve $x = \psi(t)$, $y = \phi(t)$ is given by t increasing from α at A to β at B. Suppose further that x and y are each one-valued monotonic functions of t. (In Fig. 1 we have let both x and y increase as t increases for convenience in the diagram.) We define a *line integral* along the curve AB as follows:

$$\int_{AB} g(x,y)\,\mathrm{d}x = \int_{t=\alpha}^{t=\beta} g\Big[\psi(t),\phi(t)\Big]\psi'(t)\,\mathrm{d}t. \quad (1)$$

Similarly,

$$\int_{AB} g(x,y)\,\mathrm{d}y = \int_{t=\alpha}^{t=\beta} g\Big[\psi(t),\phi(t)\Big]\phi'(t)\,\mathrm{d}t. \quad (2)$$

The integrals on the right of equations (1) and (2) are perfectly normal definite integrals. As an example, we evaluate $\int xy\,\mathrm{d}x$ (i) along the circular arc, centre the origin, joining the points A and B (Fig. 2), (ii) along the straight line AB (Fig. 3), where $A \equiv (2,0)$, $B \equiv (0,2)$.

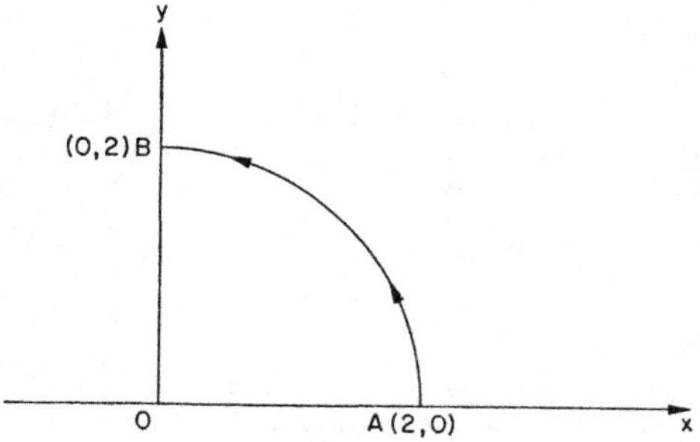

FIG. 2

(i) The parametric equations for AB are $x = 2\cos t$, $y = 2\sin t$. Hence $\mathrm{d}x = -2\sin t\,\mathrm{d}t$. As (x,y) moves along the arc AB from A to B, t changes from 0 to $\frac{1}{2}\pi$.

$$\int_{AB} xy\,\mathrm{d}x = \int_{t=0}^{t=\pi/2} 2\cos t . 2\sin t(-2\sin t)\,\mathrm{d}t$$

$$= -8\int_0^{\pi/2} \sin^2 t\cos t\,\mathrm{d}t = -8\int_{t=0}^{t=\pi/2} \sin^2 t(\mathrm{d}\sin t)$$

$$= -\frac{8}{3}\Big[\sin^3 t\Big]_0^{\pi/2}.$$

Therefore

$$\int_{AB} xy\,\mathrm{d}x = -\frac{8}{3}.$$

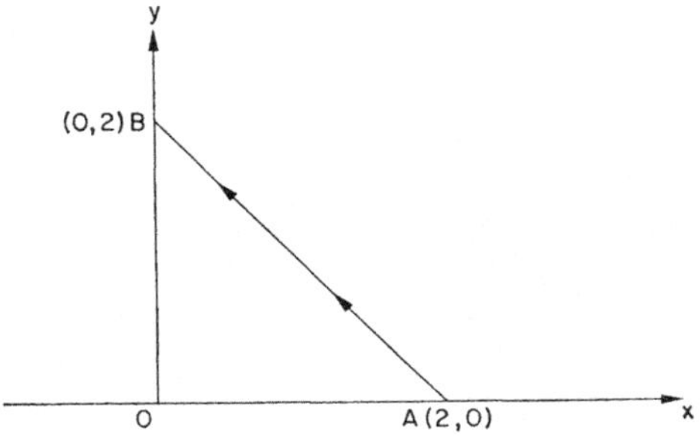

FIG. 3

(ii) The equation of AB is $x + y = 2$. So $y = 2 - x$. Writing $t = x$,

$$\int_{AB} xy\,\mathrm{d}x = \int_{t=2}^{t=0} t(2-t)\,\mathrm{d}t$$

$$= \int_{2}^{0} (2t - t^2)\,\mathrm{d}t,$$

since t changes from 2 to 0 as (x,y) moves along AB from A to B. Therefore

$$\int_{AB} xy\,\mathrm{d}x = -\frac{4}{3}.$$

Three points should be noted:

(a) The integrals in (i) and (ii) have different values. This is generally true; the value of a line integral depends on the particular path of integration AB which is chosen.

(b) $\int_{AB} g(x, y)\,\mathrm{d}x = -\int_{BA} g(x, y)\,\mathrm{d}x$; direction along the path of integration is important. The reader should check this for himself in (i) and (ii) above.

(c) In (ii) above, we could have written

$$\int_{AB} xy\,dx = \int_2^0 x(2-x)\,dx.$$

If the equation of the path is a simple relationship between x and y, there is no need to introduce a parameter t.

The definition of the line integral

$$\int_{AB} \{P(x,y)\,dx + Q(x,y)\,dy\}$$

follows at once; it is simply the sum of integrals of the types already considered.

When we have to integrate along a path on which x and y are not one-valued and monotonic, we divide the path up into sections on which these restrictions are satisfied and add the separate results. Similarly, if the path is made up of sections of different curves, we integrate along each section in turn in the appropriate direction and add the results. The following example illustrates the method.

EXAMPLE. *Evaluate* $\int\{(x^2 + y)\,dx + xy\,dy\}$ *along the path from* A *to* D *made up of the straight lines* AB, BC, CD, *where* $A \equiv (0,0)$, $B \equiv (2,2)$, $C \equiv (2,6)$, $D \equiv (3,9)$. (See Fig. 4.)

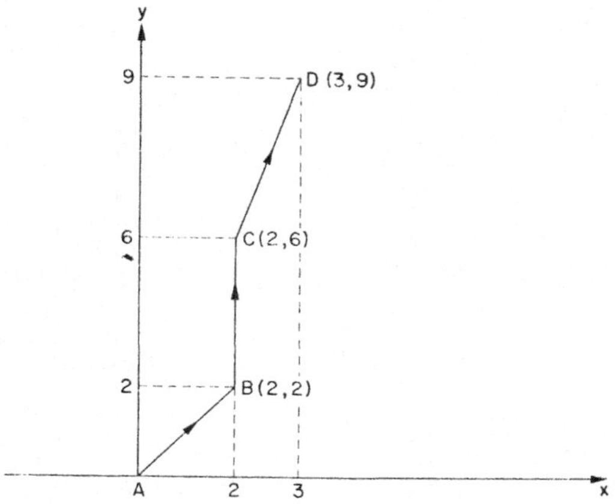

FIG. 4

The equations of the parts of the path are as follows:

AB: $y = x$; $dy = dx$.

BC: $x = 2$; $dx = 0$.

CD: $y = 3x$; $dy = 3dx$.

Write $I = \int_{ABCD} \{(x^2 + y)\,dx + xy\,dy\} = I_1 + I_2 + I_3$,

where

$$I_1 = \int_{AB} \{(x^2 + y)\,dx + xy\,dy\}$$

$$= \int_0^2 \{(x^2 + x)\,dx + x\,.\,x\,dx\}$$

$$= \int_0^2 (2x^2 + x)\,dx = 22/3,$$

$$I_2 = \int_{BC} \{(x^2 + y)\,dx + xy\,dy\} = \int_2^6 2y\,dy = 32,$$

$$I_3 = \int_{CD} \{(x^2 + y)\,dx + xy\,dy\}$$

$$= \int_2^3 \{(x^2 + 3x)\,dx + x\,.\,3x\,.\,3\,dx\}$$

$$= \int_2^3 (10x^2 + 3x)\,dx = 425/6.$$

So $$I = I_1 + I_2 + I_3 = 661/6.$$

Exercises 3:1

1. If O, A, B are the points $(0, 0)$, $(a, 0)$, $(a, 2a)$ respectively, evaluate the line integral

$$\int_O^B \{(3xy^2 + y^3)\,dx + (x^3 + 3xy^2)\,dy\}$$

when the path of integration is

(i) along the straight line OB,

(ii) along OA and then along AB,

(iii) along the arc of the parabola $y^2 = 4ax$ between O and B.

2. Evaluate $\int(x^2dy - y^2\,dx)$ between the points $(b + a, 0)$ and $(b - a, 0)$ along the arc of the circle $(x - b)^2 + y^2 = a^2$ for which $y \geqq 0$.
3. Evaluate

$$\int_C \left\{\frac{x\,dy + y\,dx}{x^2 + y^2}\right\},$$

where C consists of the part of the x-axis from plus infinity to the point $(a, 0)$, then the arc of the circle $x^2 + y^2 = a^2$ in the first, second and third quadrants from $(a, 0)$ to $(0, -a)$, and then the part of the y-axis from $(0, -a)$ to minus infinity.

3:2 Condition for independence of the path of integration

In general, the value of a line integral varies as the path of integration between the end-points A and B is changed. If the integral stays constant in value as the path changes, then its value can depend *only* on the values it takes at A and B, and this implies a restriction on the integrand. We consider the integral

$$I = \int_{AB} \{P(x, y)\,dx + Q(x, y)\,dy\}. \tag{1}$$

Suppose there exists a function $z(x, y)$, such that

$$\frac{\partial z}{\partial x} = P(x, y), \qquad \frac{\partial z}{\partial y} = Q(x, y),$$

both P and Q being here continuous functions of x and y. Then

$$dz = \frac{\partial z}{\partial x}dx + \frac{\partial z}{\partial y}dy = P(x, y)\,dx + Q(x, y)\,dy.$$

So

$$I = \int_{AB} \{P(x, y)\,dx + Q(x, y)\,dy\} = \int_{AB} dz$$

$$= [z]_{z_B}^{z_A}.$$

Therefore

$$I = z_B - z_A.$$

Hence I depends only on the values of z at A and B and so is independent of the path of integration, if there exists a function $z(x, y)$ such that $\partial z/\partial x = P(x, y)$, $\partial z/\partial y = Q(x, y)$. Now, if

$$\frac{\partial z}{\partial x} = P(x, y), \quad \frac{\partial z}{\partial y} = Q(x, y),$$

then
$$\frac{\partial^2 z}{\partial x\, \partial y} = \frac{\partial Q}{\partial x}, \quad \frac{\partial^2 z}{\partial y\, \partial x} = \frac{\partial P}{\partial y}, \quad \text{i.e. } \frac{\partial Q}{\partial x} = \frac{\partial P}{\partial y}. \tag{3}$$

So $\partial Q/\partial x = \partial P/\partial y$ is a necessary condition for independence of the path of integration.

We can also show that this condition is sufficient; if we define

$$f(x, y) = \int P(x, y)\, \mathrm{d}x,$$

where y is kept constant during the integration, then

$$\frac{\partial f}{\partial x} = P(x, y), \quad \text{and} \quad \frac{\partial^2 f}{\partial x\, \partial y} = \frac{\partial^2 f}{\partial y\, \partial x} = \frac{\partial P}{\partial y} = \frac{\partial Q}{\partial x}.$$

So
$$\frac{\partial Q}{\partial x} - \frac{\partial^2 f}{\partial x\, \partial y} = 0, \quad \text{or} \quad \frac{\partial}{\partial x}\left(Q - \frac{\partial f}{\partial y}\right) = 0.$$

This means that $(Q - \partial f/\partial y)$ is independent of x. Define

$$\phi(y) = Q - \frac{\partial f}{\partial y}. \quad \text{Then } Q = \frac{\partial f}{\partial y} + \phi(y).$$

Write
$$z(x, y) = f(x, y) + \int \phi(y)\, \mathrm{d}y.$$

This gives
$$\frac{\partial z}{\partial y} = \frac{\partial f}{\partial y} + \phi(y) = Q(x, y). \tag{4}$$

Also,
$$P(x, y) = \frac{\partial f}{\partial x} = \frac{\partial z}{\partial x}, \tag{5}$$

since $f(x,y)$, $z(x,y,)$ differ only by a function of y. So

$$\begin{aligned} P(x, y)\, \mathrm{d}x + Q(x, y)\, \mathrm{d}y &= \frac{\partial z}{\partial x}\mathrm{d}x + \frac{\partial z}{\partial y}\mathrm{d}y \\ &= \mathrm{d}z. \end{aligned}$$

Thus $$\int_{AB} \{P(x, y)\,dx + Q(x, y)\,dy\} = \int_{AB} dz$$
$$= z_B - z_A.$$

The condition (3) is therefore sufficient as well as necessary.

There are two methods of evaluating an integral which satisfies the condition (3); we may either find the function $z(x, y)$ and use (2) above, or we may choose the simplest path between the points A and B and integrate along it. We illustrate both methods in the following example: show that the line integral $\int\{(y^2 + 4)\,dx + (2xy + 3)\,dy\}$ taken between the points $A \equiv (1, 2)$ and $B \equiv (3, 4)$ is independent of the path of integration and find its value.

Comparing with (1) we have $P = y^2 + 4$, $Q = 2xy + 3$.

Therefore $$\frac{\partial P}{\partial y} = 2y = \frac{\partial Q}{\partial x}.$$

So $z(x, y)$ exists such that $\partial z/\partial x = P$, $\partial z/\partial y = Q$.

Thus $$\frac{\partial z}{\partial x} = y^2 + 4,$$

so $$z = (y^2 + 4)x + f(y),$$

where $f(y)$ is an arbitrary function of y. This gives
$$\frac{\partial z}{\partial y} = 2xy + f'(y).$$

But we know that $\partial z/\partial y = Q = 2xy + 3$. So
$$f'(y) = 3,$$

giving $f(y) = 3y + c$, where c is a constant. So
$$z = y^2x + 4x + 3y + c.$$

Using (2),
$$\int_{AB} \{P(x, y)\,dx + Q(x, y)\,dy\} = \left[xy^2 + 4x + 3y + c\right]_A^B = 58.$$

To integrate along a suitable path between A and B we choose that path which produces the simplest possible integration; this is not necessarily the simplest path geometrically. Here we choose a path made up of straight lines AC, CB as shown in Fig. 5.

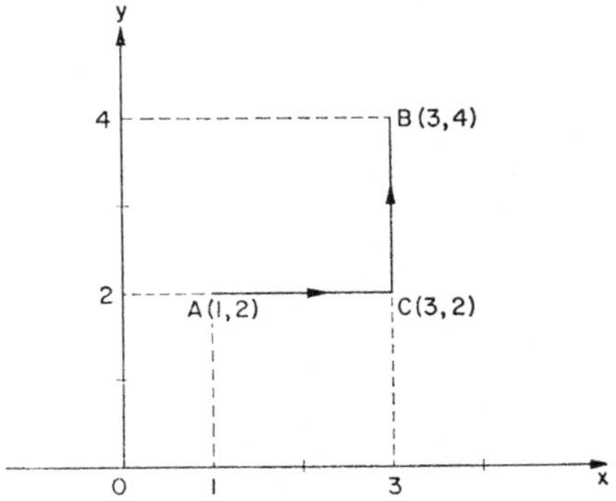

FIG. 5

On AC, $y = 2$; thus $dy = 0$.
On CB, $x = 3$; thus $dx = 0$.

So, $$\int_A^B \{P(x, y)\,dx + Q(x, y)\,dy\}$$

$$= \int_{AC} (P\,dx + Q\,dy) + \int_{CB} (P\,dx + Q\,dy)$$

$$= \int_1^3 8\,dx + \int_2^4 (6y + 3)\,dy$$

$$= 58.$$

Exercises 3:2

1. Show that

$$\int_{(1,1)}^{(2,3)} \left[\frac{2x - y}{x^2 + y^2}\,dx + \frac{2y + x}{x^2 + y^2}\,dy\right]$$

is independent of the path of integration between the points (1, 1), (2, 3), and find its value.

2. Determine the constant α so that the line integral

$$\int\left[\frac{1+\alpha y^2}{(1+xy)^2}\,\mathrm{d}x+\frac{1+\alpha x^2}{(1+xy)^2}\,\mathrm{d}y\right]$$

taken between two fixed points shall be independent of the path of integration.
Evaluate the integral with the *general* value of α
(i) from (0, 0) to (1, 0) and thence to (1, 1) by straight lines,
(ii) from (0, 0) to (1, 1) by a single straight-line path.
Using these results, verify the value of α obtained for independence of the integral on the path. [B.Sc.(Eng.)]

3. Show that a function $z(x, y)$ exists such that

$$\mathrm{d}z=\left(2xy+\frac{y}{1+x^2}+\sin^2\pi y\right)\mathrm{d}x+(x^2+\tan^{-1}x+\pi x\sin 2\pi y)\,\mathrm{d}y.$$

Hence prove that

$$\int_{(0,0)}^{(1,2)}\left\{\left(2xy+\frac{y}{1+x^2}+\sin^2\pi y\right)\mathrm{d}x+\left(x^2+\tan^{-1}x+\pi x\sin 2\pi y\right)\mathrm{d}y\right\}=2(1+\tfrac{1}{4}\pi).$$

[B.Sc.(Eng.)]

3:3 Another type of line integral

We can also define line integrals of the form $\int_{AB} g(x, y)\,\mathrm{d}s$, where s denotes arc-length on the path of integration. For if the path is given parametrically by the equations $x = \psi(t)$, $y = \phi(t)$, then

$$\left(\frac{\mathrm{d}s}{\mathrm{d}t}\right)^2=\left(\frac{\mathrm{d}x}{\mathrm{d}t}\right)^2+\left(\frac{\mathrm{d}y}{\mathrm{d}t}\right)^2=\left(\frac{\mathrm{d}\psi}{\mathrm{d}t}\right)^2+\left(\frac{\mathrm{d}\phi}{\mathrm{d}t}\right)^2.$$

(With our previous restriction on $\psi(t)$, $\phi(t)$ that they be monotonic functions, then $s(t)$ is a monotonic increasing function.)

So
$$\int_{AB} g(x, y)\,\mathrm{d}s=\int_{t=\alpha}^{t=\beta} g[\psi(t),\phi(t)]\sqrt{\left\{\left(\frac{\mathrm{d}\psi}{\mathrm{d}t}\right)^2+\left(\frac{\mathrm{d}\phi}{\mathrm{d}t}\right)^2\right\}}\,\mathrm{d}t. \quad (1)$$

It may be simpler in some cases to integrate in terms of x. Then

$$\int_{AB} g(x, y)\,\mathrm{d}s=\int_{x_1}^{x_2} g(x, y)\sqrt{\left\{1+\left(\frac{\mathrm{d}y}{\mathrm{d}x}\right)^2\right\}}\,\mathrm{d}x, \quad (2)$$

where y is now a function of x, and x_1, x_2 are the values of x at A and B respectively. Here, care must be taken to attach the correct sign to the expression for $\mathrm{d}s$ throughout the path, so that s is monotonic increasing always.

For a curve given in polar coordinates by an equation of the form $\rho = f(\psi)$, we can evaluate integrals such as $\int_{AB} g(\rho, \psi)\,\mathrm{d}s$. Now

$$\int_{AB} g(\rho, \psi)\,\mathrm{d}s = \int_{\psi_1}^{\psi_2} g[f(\psi), \psi]\sqrt{\left\{\rho^2 + \left(\frac{\mathrm{d}\rho}{\mathrm{d}\psi}\right)^2\right\}}\,\mathrm{d}\psi, \tag{3}$$

where $\psi = \psi_1$, $\psi = \psi_2$ correspond to A, B respectively.

The path of integration is often a closed curve. Normally the direction of integration is anti-clockwise round the curve, the conventional indication of this being $\oint_C$, although sometimes $\oint$ is used. If the closed curve is denoted by C, then C takes the place of AB below the integral sign.

As an example, consider $\oint_C (x^4 + y^4)\,\mathrm{d}s$, where C is the boundary of the circle $x^2 + y^2 = a^2$.

On the circle we can write $x = a\cos t$, $y = a\sin t$, and the path is traced out as t goes from 0 to 2π. In this case

$$\left(\frac{\mathrm{d}s}{\mathrm{d}t}\right)^2 = \left(\frac{\mathrm{d}x}{\mathrm{d}t}\right)^2 + \left(\frac{\mathrm{d}y}{\mathrm{d}t}\right)^2 = a^2,$$

yielding

$$\mathrm{d}s = a\,\mathrm{d}t,$$

taking s to increase with t throughout the path.

So

$$\oint_C (x^4 + y^4)\,\mathrm{d}s = \int_0^{2\pi} (a^4\cos^4 t + a^4\sin^4 t)\,a\,\mathrm{d}t$$

$$= a^5\int_0^{2\pi} (\cos^4 t + \sin^4 t)\,\mathrm{d}t$$

$$= 4a^5\int_0^{\pi/2} (\cos^4 t + \sin^4 t)\,\mathrm{d}t$$

$$= 4a^5 \,.\, 2 \,.\, \frac{3}{4} \cdot \frac{1}{2} \cdot \frac{\pi}{2},$$

since
$$\int_0^{\pi/2} \cos^4 t\,dt = \int_0^{\pi/2} \sin^4 t\,dt.$$

So
$$\oint_C (x^4 + y^4)\,ds = \tfrac{3}{2}\pi a^5.$$

Exercises 3:3

1. Evaluate $\int (1/\rho)\,ds$ taken over the arc of the lemniscate $\rho^2 = a^2 \cos 2\psi$ for which $0 \leqq \psi \leqq \frac{1}{6}\pi$, ds denoting arc-length.

2. Evaluate $\oint (4r - 3a)^{-3/2}\,ds$ taken round the whole perimeter of the curve $r = a(\cos\theta + 2)$. [B.Sc.(Eng.)]

3:4 Line integrals in three dimensions

The results of the previous sections can easily be extended to three dimensions. We thus have line integrals such as

$$\int_{AB} \{P(x, y, z)\,dx + Q(x, y, z)\,dy + R(x, y, z)\,dz\} \tag{1}$$

and
$$\int_{AB} F(x, y, z)\,ds, \tag{2}$$

where, in (2), $ds^2 = dx^2 + dy^2 + dz^2$. We can also integrate round a closed curve in three dimensions.

Here, too, the value of a line integral generally depends on the path chosen. In order that the integral in (1) shall be independent of the path, there must exist a function $w(x, y, z)$ such that $\partial w/\partial x = P$, $\partial w/\partial y = Q$, $\partial w/\partial z = R$. If this is so, then $\partial P/\partial y = \partial^2 w/\partial y\,\partial x = \partial Q/\partial x$, with two similar results. So a necessary condition for independence of the path is

$$\frac{\partial P}{\partial y} = \frac{\partial Q}{\partial x}, \quad \frac{\partial Q}{\partial z} = \frac{\partial R}{\partial y}, \quad \frac{\partial R}{\partial x} = \frac{\partial P}{\partial z}. \tag{3}$$

This condition can also be shown to be sufficient.

When this condition is satisfied, it is usually easy to evaluate the integral by choosing a path between A and B made up of straight lines parallel to the coordinate axes. For example

$$\int_{AB} \{(2xy + z)\,dx + x^2\,dy + (3z^2 + x)\,dz\}$$

is independent of the path of integration, since it clearly satisfies condition (3). If A is the point (1, 1, 1) and B is the point (2, 3, 4), a suitable path is along the straight lines AC, CD, DB, where C is the point (2, 1, 1) and D is (2, 3, 1) (see Fig. 6).

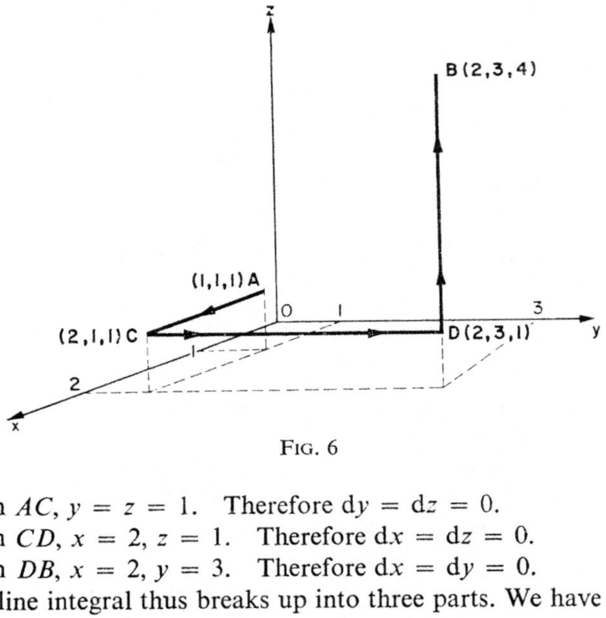

FIG. 6

On AC, $y = z = 1$. Therefore $dy = dz = 0$.
On CD, $x = 2$, $z = 1$. Therefore $dx = dz = 0$.
On DB, $x = 2$, $y = 3$. Therefore $dx = dy = 0$.
The line integral thus breaks up into three parts. We have

$$\int_{AB} \{(2xy + z)\,dx + x^2\,dy + (3z^2 + x)\,dz\}$$

$$= \int_{AC} (2x + 1)\,dx + \int_{CD} 4\,dy + \int_{DB} (3z^2 + 2)\,dz$$

$$= \int_1^2 (2x + 1)\,dx + \int_1^3 4\,dy + \int_1^4 (3z^2 + 2)\,dz$$

$$= [x^2 + x]_1^2 + [4y]_1^3 + [z^3 + 2z]_1^4$$

$$= 81.$$

Alternatively, we can find the function $w(x, y, z)$. We have $\partial w/\partial x = 2xy + z$; so, integrating with respect to x, $w = x^2y + + 2x + f(y, z)$, where $f(y, z)$ is an arbitrary function of y and z. This value of w implies that $\partial w/\partial y = x^2 + \partial f/\partial y$; but we must have $\partial w/\partial y = Q = x^2$. So $\partial f/\partial y = 0$; that is $f(y, z)$ is in fact independent of y. Similarly, comparing $\partial w/\partial z = x + \partial f/\partial z$ with $R = 3z^2 + x$, we find that $\partial f/\partial z = 3z^2$. Thus $f = z^3 + c$, where c is an arbitrary constant. Then

$$w(x, y, z) = x^2y + zx + z^3 + c.$$

The value of the line integral is $w(2, 3, 4) - w(1, 1, 1)$, and on evaluation this gives 81, as before.

The evaluation, in three dimensions, of line integrals which are not independent of the path is generally easiest when x, y, z are given on the path of integration in terms of a parameter.

Line integrals play an important part in Vector Field Theory, and we shall return to them later in Chapters 6 and 7.

Exercises 3:4

1. Find the value of $\int(y^2\,dx + xy\,dy + zx\,dz)$ between the points (0, 0, 0), (1, 1, 1) when the path is
   (i) the straight line joining (0, 0, 0), (1, 1, 1),
   (ii) along the x-axis, then parallel to the y-axis, then parallel to the z-axis,
   (iii) along the curve $x = t$, $y = t^2$, $z = t^3$.
2. Show that a function $w(x, y, z)$ exists such that

$$dw = (2xy + 3z^2)\,dx + (x^2 + 4yz)\,dy + (2y^2 + 6xz)\,dz,$$

and find the expression for w in terms of x, y, z. Hence evaluate the line integral

$$\int\{(2xy + 3z^2)\,dx + (x^2 + 4yz)\,dy + (2y^2 + 6yz)\,dz\}$$

along any path joining the points (0, 0, 0) and (1, 1, 1). Check your result by direct integration in which the path is the straight line $x = y = z$.

Miscellaneous Exercises 3

1. Evaluate $\oint_C \{(x^2 + 2y)\,\mathrm{d}x + (y - x)\,\mathrm{d}y\}$ where C is the closed curve made up of segments of the straight lines $x = 0$, $y = 0$, $y = \frac{1}{2}x + 1$, and part of the parabola $y = \frac{1}{2}(x - 4)^2$.
2. Show that $\oint_C (x^2y\,\mathrm{d}x + x^3\,\mathrm{d}y) = \frac{1}{8}\pi$, where C is the curve made up of the straight line joining (0, 0) to (1, 0), the arc of the circle $x^2 + y^2 = 1$ which lies in the positive quadrant, and the straight line joining (0, 1) to (0, 0).
3. Find the value of the line integral

$$\int \{(x^2 + yz)\,\mathrm{d}x + (z^2 - x^2)\,\mathrm{d}y + xy\,\mathrm{d}z\},$$

from the point $O(0, 0, 0)$ to $A(1, 1, 1)$ when taken along

(i) $OBCA$, where OB is along the x-axis and BC, CA are parallel to the y- and z-axes respectively,

(ii) the curve $x = 1 - \cos\theta$, $y = \sin\theta$, $z = 2\theta/\pi$.

4. Show that the line integral

$$\int_{(1,1,1)}^{(2,3,4)} \{(x^2 - yz)\,\mathrm{d}x + (y^2 - zx)\,\mathrm{d}y + (z^2 - xy)\,\mathrm{d}z\}$$

is independent of the path of integration between the end-points (1, 1, 1), (2, 3, 4), and find its value.

5. Determine the constants a, b so that

$$(y + a\,\mathrm{e}^{2x}\cos y + \sin^2 y)\,\mathrm{d}x + (x - \mathrm{e}^{2x}\sin y + bx\sin 2y)\,\mathrm{d}y$$

is the exact differential of a function $u(x, y)$ and find $u(x, y)$. Hence, with these values of a and b, find the value of the line integral

$$\int \{(y + a\,\mathrm{e}^{2x}\cos y + \sin^2 y)\,\mathrm{d}x + (x - \mathrm{e}^{2x}\sin y + bx\sin 2y)\,\mathrm{d}y\}$$

taken along any path between the points $(1, \frac{1}{2}\pi)$, $(4, \pi)$.

6. A particle moves along the curve defined by the equation $x^2 = 4y$, $3x^3 = 8z$ under the action of a force whose components in the directions of the coordinate axes are $3x^2$, $2xz - y$ and $2z$ units respectively. The work done in moving the particle from a point A to another point B on the curve is given by the line integral

$$\int_A^B \{(3x^2\,\mathrm{d}x + (2xz - y)\,\mathrm{d}y + 2z\,\mathrm{d}z)\}.$$

Find the work done when A is the origin and B the point (4, 4, 24).

7. The closed curve C has equations $x = a\cos^3\theta$, $y = a\sin^3\theta$.
Evaluate (i) $\oint_C (3x^2 - 2y^2)\,ds$, s denoting arc-length,
(ii) $\oint_C (x\,dy - y\,dx)$.
Check by direct integration that the value given by (ii) is equal to twice the area enclosed by C.
[In (i), care must be taken to see that $ds/d\theta$ is positive at all points of C.]

8. (i) Evaluate the line integral $\int (1/\rho^3)\,ds$, taken along the spiral $\rho = a\,e^{\psi}$ from $\psi = -4\pi$ to $\psi = 4\pi$.
(ii) Evaluate the line integrals
(a) $\int (x^2 + y^2 - z^2)\,ds$, s denoting arc-length,
(b) $\int \{12xz\,dx - 6y\,dy + (1 + 6x^2)\,dz\}$,
taken along the circular helix $x = a\cos t$, $y = a\sin t$, $z = at$ from $t = 0$ to $t = 2\pi$.

9. (a) Show that the line integral

$$\int \{yz/(x^2 + y^2)\}\,ds$$

taken along the arc of the space-curve given by the equations $x = 3at$, $y = 3at^2$, $z = 2a(1 + t^3)$ between the points for which $t = 0$ and $t = 1$, has the value $a(4 - \frac{1}{2}\pi + \log 2)$.
(b) Find the value of the integral

$$\int \{(y + 3z)\,dx + (2z + x)\,dy + (3x + 2y)\,dz\}$$

taken along the arc of the helix $x = a\cos\theta$, $y = a\sin\theta$, $z = 2a\theta/\pi$ between the points $(a, 0, 0)$, $(0, a, a)$. [B.Sc.(Eng.)]

CHAPTER 4

MULTIPLE INTEGRALS

4:1 An integral expression for the volume of a cylinder

We consider now how to express the volume of a certain solid by means of an integral. Suppose C is a simple closed curve in the x–y plane, such that no line parallel to either axis meets it more than twice. Denote by A the area enclosed by C. Then the element of area dA corresponding to the linear elements dx, dy

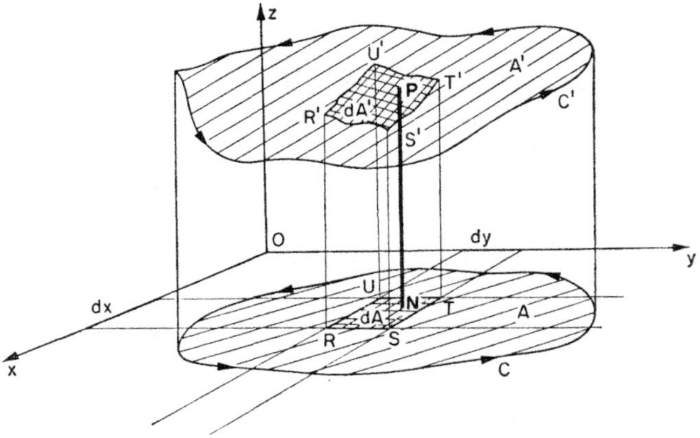

FIG. 7

is $dA = dx\ dy$. Suppose that $z = f(x, y)$ is the equation of a surface in three dimensions, such that to each point (x,y) in the x–y plane corresponds a single point (x,y,z) on the surface. Imagine a line parallel to the z-axis moving continuously round the curve C. It will cut the surface $z = f(x,y)$ in a curve C' (of which C is the projection in the x–y plane). When C has been completely traversed we shall have a volume in the form of a cylinder with generators parallel to Oz, enclosed at the bottom by A and at the top by a surface A' which is part of the surface $z = f(x,y)$, as in Fig. 7. We note that whereas A is a plane area, the area A' is in general part of a curved surface, and will be a plane area only if $z = f(x,y)$ is the equation of a plane. We denote the volume of the cylinder by V, and look for an integral expression for V.

Consider the element of area $\mathrm{d}A = \mathrm{d}x\,\mathrm{d}y$. At any point N of $\mathrm{d}A$ draw a line NP parallel to Oz meeting A' at P. Denoting the vertices of $\mathrm{d}A$ by R, S, T, U, there is a corresponding area $R'S'T'U'$ on A'. An approximate value for the volume standing on $\mathrm{d}A$ as base is $NP.\,\mathrm{d}A = z\,\mathrm{d}x\,\mathrm{d}y$, where z is the value of $f(x,y)$ at N. The smaller we make $\mathrm{d}x\,\mathrm{d}y$, the better this approximation becomes to the element of volume defined by $RSTUR'S'T'U'$. In fact,

$$V = \lim_{\mathrm{d}A \to 0} \sum z\,\mathrm{d}A,$$

where Σ denotes summation over all possible elements $\mathrm{d}A$ within A. We write this limit as

$$V = \iint_A z\,\mathrm{d}A = \iint_A f(x,y)\,\mathrm{d}A, \tag{1}$$

and call it the double integral of $f(x,y)$ over the area A.

Note that in (1) z is a function of x and y. We now have to find some way of evaluating (1) when z and A are given.

4:2 Evaluation of $\iint_A z\,\mathrm{d}A$

We consider now an alternative way of finding an expression for V. Suppose the extreme values of x on C are a, b, where $a < b$ (see Fig. 8). Take a plane $DEFG$ parallel to the plane yOz, cutting Ox at a distance x from O. This gives a plane section $DEFG$ of the volume V, FG being in general a curve on the surface A'. Suppose the area of $DEFG$ is $\phi(x)$, an assumption which we shall justify later. Take a plane parallel to $DEFG$ distant $x + \mathrm{d}x$ from O. Then the element of volume between

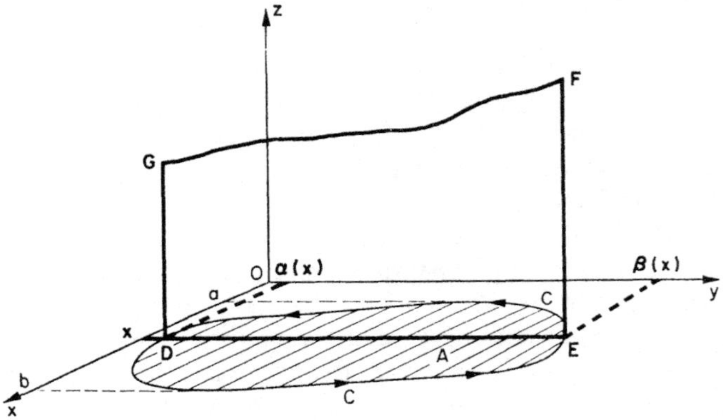

FIG. 8

the two planes is $\phi(x)\,\mathrm{d}x$ to a first approximation. (This element is in fact a slice of V.) So

$$V = \int_a^b \phi(x)\,\mathrm{d}x. \tag{1}$$

We have now to find $\phi(x)$: let the values of y at D, E be α, β respectively. The curve C in the x–y plane has an equation of the form $g(x,y) = 0$, so y is a function of x on C. Hence α, β are functions of x. Further, GF is a curve on which x remains constant, and only y and z vary. The area $DEFG$ we require is the area *under* the curve GF, as y varies from α to β. This area can thus be found in the normal way by means of a definite integral. In fact

$$DEFG = \int_\alpha^\beta z(x, y)\,\mathrm{d}y. \tag{2}$$

In (2), x is constant, and y varies. So $\int z(x, y)\,\mathrm{d}y$ is simply an integral of a function of y with respect to y. Once the integration has been performed, the limits α, β for y are put in; α, β are themselves functions of x, so the resulting expression for $DEFG$ is a function of x, and we are justified in writing

$$DEFG = \phi(x) = \int_{\alpha}^{\beta} z(x, y)\, \mathrm{d}y.$$

Substituting in (1) we obtain

$$V = \int_{a}^{b} \left[\int_{\alpha}^{\beta} z(x, y)\, \mathrm{d}y \right] \mathrm{d}x. \tag{3}$$

We have thus obtained an alternative expression for V. It enables us to evaluate the double integral $\iint_A z\, \mathrm{d}A$ of § 4:1 by means of a repeated integral. In theory, the function $z = f(x,y)$ or the bounding curve C of the area A may be so defined that the double integral is not equal to the corresponding repeated integral; see W. L. Ferrar, *Integral Calculus*, Oxford University Press, 1958, pp. 163–74. In all the integrals we shall meet, it will always be possible to convert a double integral to a repeated integral.

Relation (3) above is written either as

$$V = \int_{a}^{b} \int_{\alpha}^{\beta} z\, \mathrm{d}y\, \mathrm{d}x, \tag{4}$$

or as

$$V = \int_{a}^{b} \mathrm{d}x \int_{\alpha}^{\beta} z\, \mathrm{d}y. \tag{5}$$

The reader should note that in (4) since α, β are functions of x, the *first* integration must be with respect to y, giving a function of x to be integrated with respect to x between limits a and b, which are constants. So the right-hand side of (4) is more precisely written

$$\int_{a}^{b} \left[\int_{\alpha}^{\beta} z\, \mathrm{d}y \right] \mathrm{d}x$$

as in (3), but in practice the brackets are omitted. The reader may also meet examples of double integration in which the $\mathrm{d}x$ and $\mathrm{d}y$ of (4) are interchanged although the same order of integration is to be followed. In general, this causes no difficulty; if α, β are functions of x, then clearly the first integration must be with respect to y, whether the double integral is written as

$$\int_a^b \int_\alpha^\beta z\, \mathrm{d}y\, \mathrm{d}x, \quad \text{or as} \quad \int_a^b \int_\alpha^\beta z\, \mathrm{d}x\, \mathrm{d}y.$$

Ambiguity could arise only in the special case where α, β are *both*, in fact, constants. In the following pages we shall use the order of $\mathrm{d}x$ and $\mathrm{d}y$ given in (4), i.e. the inner integral and the inner "d" sign belong together.

Clearly we could also find the value of V by dividing the volume up into slices parallel to the plane xOz. We should then obtain

$$V = \int_{a'}^{b'} \int_{\alpha'}^{\beta'} z(x, y)\, \mathrm{d}x\, \mathrm{d}y,$$

where α', β' are now functions of y and the first integration is with respect to x. The second integration is with respect to y between constant limits a' and b'. It follows that we can integrate in either order; the *value* of the integral is independent of the order of integration, but the *limits of integration* do depend on it. We shall return to this point in § 4:3.

Double integrals have a very useful physical application in finding volumes, and we have used this to introduce them. However, the double integral $\iint f(x,y)\, \mathrm{d}x\, \mathrm{d}y$ is a mathematical entity which does not need a physical interpretation any more than $\int y\, \mathrm{d}x$ does. The point to note here is that to evaluate $\iint_A f(x,y)\, \mathrm{d}A$ or, as it is often written, $\iint_A f(x,y)\, \mathrm{d}x\, \mathrm{d}y$, over an area A, we convert it to the form

$$\int_a^b \int_\alpha^\beta f(x, y)\, \mathrm{d}y\, \mathrm{d}x \quad \text{or} \quad \int_{a'}^{b'} \int_{\alpha'}^{\beta'} f(x, y)\, \mathrm{d}x\, \mathrm{d}y,$$

whichever is the more convenient, the limits being determined by the form of A and its boundary curve C.

The standard method of dealing with the evaluation of $\iint_A f(x,y)\,\mathrm{d}x\,\mathrm{d}y$ involves

(i) a sketch of the area A,
(ii) a decision as to the order of integration,
(iii) the determination of the limits,
(iv) the integration itself.

We illustrate this by means of an example; evaluate $\iint 2xy\,\mathrm{d}A$ over the area A bounded by $y^2 = x$, $x^2 = y$, $x = \frac{1}{2}$, for which $x \leqq \frac{1}{2}$.

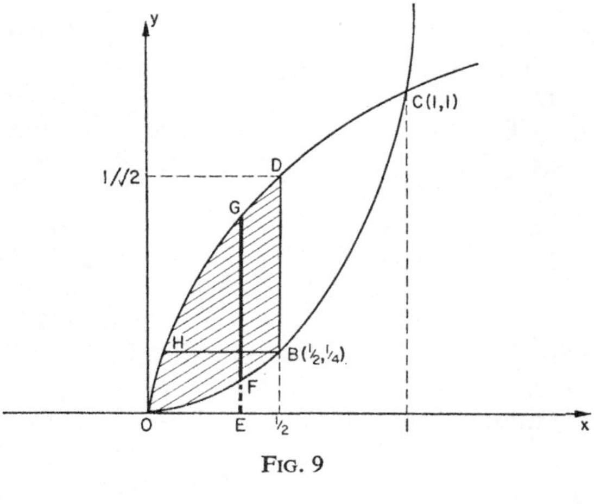

FIG. 9

The parabolas $y^2 = x$, $x^2 = y$ meet at the origin and at the point C (1, 1). The area of integration is OBD, where B is the point (1/2, 1/4) and D the point (1/2, $1/\sqrt{2}$). Let us try integrating first with respect to y. We thus have to determine limits for y as functions of x. The range of variation of x is clearly from 0 to $\frac{1}{2}$. If we keep x constant within this range, say at E (Fig. 9), then y varies from its value at F to its value at G; that is from $y = x^2$ at

F to $y = \sqrt{x}$ at G. Hence

$$\iint_A 2xy\,\mathrm{d}A = \iint_A 2xy\,\mathrm{d}y\,\mathrm{d}x$$
$$= \int_0^{1/2} \mathrm{d}x \int_{x^2}^{\sqrt{x}} 2xy\,\mathrm{d}y$$
$$= \int_0^{1/2} x[y^2]_{x^2}^{\sqrt{x}}\,\mathrm{d}x$$
$$= \int_0^{1/2} x[x - x^4]\,\mathrm{d}x$$
$$= \left[\frac{x^3}{3} - \frac{x^6}{6}\right]_0^{1/2}$$
$$= 5/128.$$

We know that we must get the same result by integrating first with respect to x. This involves keeping y constant to find limits for x as functions of y, and then letting y vary from 0 to $1/\sqrt{2}$. The reader will see from Fig. 9 that the same limits for x are not valid over the whole range of y. We have to split the range for y and write

$$\iint_A 2xy\,\mathrm{d}x\,\mathrm{d}y = \int_0^{1/4} dy \int_{y^2}^{y} 2xy\,\mathrm{d}x + \int_{1/4}^{1/\sqrt{2}} \mathrm{d}y \int_{y^2}^{1/2} 2xy\,\mathrm{d}x,$$

the integrals on the right corresponding to the areas OBH, HBD respectively. This method is needlessly long: it is important to determine the simpler order of integration.

It often happens that we have to evaluate a double integral of the form

$$\int_{a_1}^{a_2} \mathrm{d}x \int_{b_1}^{b_2} g(x)h(y)\,\mathrm{d}y$$

where a_1, a_2, b_1, b_2 are all constants, and $f(x, y)$ of the theory is replaced by the product $g(x)h(y)$. When this is so, the integral is is equal to the product

$$\int_{a_1}^{a_2} g(x)\,\mathrm{d}x \times \int_{b_1}^{b_2} h(y)\,\mathrm{d}y$$

of two simple integrals.

The area over which we are required to integrate is frequently called the field of integration.

The reader is strongly recommended to work the Exercises which follow, to gain facility in the basic methods of evaluation of double integrals.

Exercises 4:2

1. Show that the value of $\iint e^{x-y}\,dx\,dy$ taken over the triangular area enclosed by the coordinate axes and the line $x + y = \log a$, $(a > 1)$, is $(a - 1)^2/2a$. [B.Sc.(Eng.)]
2. Evaluate the integral $\iint x^2y\,dx\,dy$ when the field of integration is the area enclosed by the curves $y^2 = 8x$, $x^2 = y$.
3. Evaluate $\iint \cos\{\pi(x + y)\}\,dx\,dy$ over the area in the first quadrant bounded by the axes and the line $x + y = 1$. Find also the value of the integral over the area bounded by the axes and the lines $x + y = 1$, $x + y = 2$.
4. Evaluate $\iint(\cos x - 1)\,dx\,dy$ over the triangular area bounded by the lines $y = 2x/\pi$, $x = \frac{1}{2}\pi$, $y = 0$.
5. Evaluate
$$\iint \frac{x}{y\sqrt{(y - x^2)}}\,dx\,dy$$
over the area in the first quadrant bounded by $y = x^2$, $y = x$.
6. Evaluate $\iint[x/(x + y)]\,dy\,dx$ over the area of the triangle whose vertices are at the points $(a, 0)$, $(a, 2a)$, $(0, a)$. [B.Sc.(Eng.)]
7. Evaluate $\iint x^2y\,dy\,dx$ over the closed region of the x–y plane bounded by the lines $x = 0$, $x = y$, $x = 2 - y$.

4:3 Change of order of integration

It frequently happens that a given double integral is in such a form that integration in the order given is impracticable if not impossible. In such cases we must *change* the order of integration. The method is best illustrated by a specific example. Consider the integral I, where

$$I = \int_0^a dx \int_0^x \frac{\sec^2 y}{\sqrt{\{(a - x)(x - y)\}}}\,dy \quad (a > 0).$$

As given, the first integration is with respect to y, the factor

$\sqrt{(a - x)}$ in the denominator being treated as a constant at this stage. Attempts to perform the integration as given will soon convince the reader that it is desirable to change the order and integrate first with respect to x, for then the first integration involves

$$\int \frac{dx}{\sqrt{\{(a-x)(x-y)\}}},$$

with y treated as a constant; this integral can easily be found. We begin the change of order by sketching the area of integration. From the form of I, we see that the limits for x are 0 and a, while for x held constant between 0 and a, y varies from $y = 0$ to $y = x$. Hence the area of integration is the triangle OAB in Fig. 10.

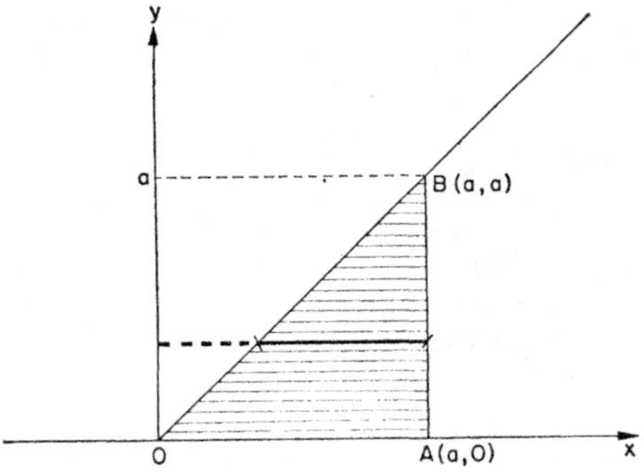

FIG. 10

We can now obtain the limits for integration in the reversed order. From the figure, the limits for y are 0 and a, while for y held constant between 0 and a, x varies from its value on OB to its value on AB; that is from $x = y$ to $x = a$. Hence

$$I = \int_0^a dy \int_y^a \frac{\sec^2 y}{\sqrt{\{(a-x)(x-y)\}}}\, dx.$$

So $$I = \int_0^a \sec^2 y \, dy \int_y^a \frac{dx}{\sqrt{\{(a-x)(x-y)\}}}.$$

If we put $(a - x) = (x - y)\,t^2$, we find that

$$\begin{aligned}\int \frac{dx}{\sqrt{\{(a-x)(x-y)\}}} &= \int \frac{-2\,dt}{(t^2+1)} \\ &= -2\tan^{-1}(t) \\ &= -2\tan^{-1}\sqrt{\{(a-x)/(x-y)\}}.\end{aligned}$$

Therefore $$\begin{aligned} I &= \int_0^a \sec^2 y \left[-2\tan^{-1}\sqrt{\left(\frac{a-x}{x-y}\right)}\right]_{x=y}^{x=a} dy \\ &= \int_0^a \sec^2 y (2\,.\,\pi/2)\,dy \\ &= \pi\,[\tan y]_0^a\,.\end{aligned}$$

So $$I = \pi \tan a.$$

In all cases of change of order, the method to be followed is the same: *first*, sketch the area of integration given by the limits of the integral, *then* find the limits for the new order of integration from this sketch.

Exercises 4:3

1. Change the order of integration in the integral

$$\int_0^1 dy \int_{y^2}^y (y\,e^x/x)\,dx,$$

and hence evaluate the integral.

2. Change the order of integration in

$$\int_0^1 \int_0^{(2x-2)/(x-3)} f(x, y)\,dy\,dx,$$

and evaluate this integral when $f(x, y) = x(y - 2)^2$. [B.Sc.(Eng.)]

3. Show in a diagram the areas of integration for the integrals

$$\int_0^1 \mathrm{d}x \int_0^x xy\,\mathrm{d}y, \qquad \int_1^{\sqrt{2}} \mathrm{d}x \int_0^{\sqrt{(2-x^2)}} xy\,\mathrm{d}y.$$

Change the order of integration in each, and hence express the sum of these integrals as one double integral and evaluate it.

4. (i) Evaluate

$$\int_0^a \mathrm{d}x \int_x^a \{y/\sqrt{(x^2+y^2)}\}\,\mathrm{d}y.$$

(ii) Change the order of integration in the integral in (i), evaluate the double integral so obtained, and verify that both orders of integration give the same result.

4:4 Change of variables in double integrals

Many double integrals can be greatly simplified by a change of variables. In particular, the change from cartesian coordinates x, y to plane polar coordinates ρ, ψ is most useful. We shall give now the theory of the general change of variables x, y to u, v; readers who are interested only in the practical evaluation of double integrals can omit this theory, noting the statement of the transformation to be applied in (1) below, and go on to the examples of its use at the end of this section and in § 4:5.

We consider $\iint_A f(x, y)\,\mathrm{d}x\,\mathrm{d}y$ under the transformation $x = \theta(u,v)$, $y = \phi(u,v)$, which transforms the area A of the x–y plane to an area D of the u–v plane. We suppose that $f(x,y)$ is continuous in A and on its boundary, that $\partial x/\partial u$, $\partial x/\partial v$, $\partial y/\partial u$, $\partial y/\partial v$ are continuous in D and on its boundary, and that $J = \partial(x,y)/\partial(u,v)$ and $\partial x/\partial u$ do not vanish in D. Then we shall show that

$$\iint_A f(x,y)\,\mathrm{d}x\,\mathrm{d}y = \iint_D F(u,v)\,|J|\,\mathrm{d}u\,\mathrm{d}v, \qquad (1)$$

where $F(u, v) = f\{\theta(u, v), \phi(u, v)\}$, $J = \partial(x, y)/\partial(u, v)$. Thus F is obtained by substituting for x, y in $f(x,y)$; $\mathrm{d}x\,\mathrm{d}y$ is replaced by $|J|\,\mathrm{d}u\,\mathrm{d}v$, $|J|$ being the modulus of J, and the area A is changed to the area D.

We establish (1) in two stages, by replacing the given transformation by two others:

(a) $x = \xi, \quad y = \lambda(\xi, \eta),$

(b) $\xi = \theta(u, v), \quad \eta = v,$

where $\phi(u,v)$ reduces to $\lambda(\xi, \eta)$, when u and v are expressed in terms of ξ, η by (b). Then $\phi(u, v) = \lambda[\theta(u,v), v]$.

Making use of a property of Jacobians (Exercises 2:6, No. 3),

$$J = \frac{\partial(x, y)}{\partial(u, v)} = \frac{\partial(x, y)}{\partial(\xi, \eta)} \cdot \frac{\partial(\xi, \eta)}{\partial(u, v)} = \frac{\partial \lambda}{\partial \eta} \frac{\partial \theta}{\partial u}. \tag{2}$$

Since J and $\partial\theta/\partial u (= \partial x/\partial u)$ are continuous and do not vanish in D, it follows that J, $\partial\theta/\partial u$, $\partial\lambda/\partial\eta$ are continuous and of constant sign in D.

Let Δ be the region of the ξ–η plane which corresponds to A by (a) and to D by (b). Using Fig. 11(i), we write

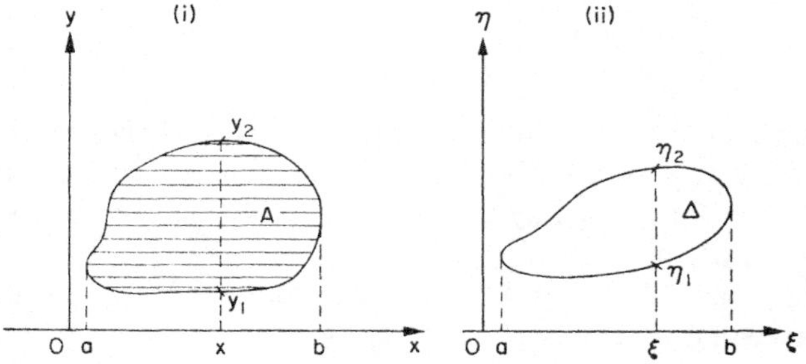

FIG. 11

$$\iint_A f(x, y)\,\mathrm{d}x\,\mathrm{d}y = \int_a^b \mathrm{d}x \int_{y_1}^{y_2} f(x, y)\,\mathrm{d}y.$$

Keep x constant, and equal to ξ. Substitute for y, using (a):

$$\iint_A f(x, y)\,\mathrm{d}x\,\mathrm{d}y = \int_a^b d\xi \int_{\eta_1}^{\eta_2} f\{\xi, \lambda(\xi, \eta)\} \left|\frac{\partial \lambda}{\partial \eta}\right| \mathrm{d}\eta. \tag{3}$$

In (3), dy has been replaced by $|\partial\lambda/\partial\eta|\,d\eta$. This is the normal change of variable in integration. For if $y = g(t)$ say, then $dy = (dg/dt)\,dt$. Here $y = \lambda(\xi,\eta)$, with ξ constant, so that y is a function of η, and $dy/d\eta = \partial\lambda/\partial\eta$. If now $\partial\lambda/\partial\eta > 0$, then y increases with η, and y_1 corresponds to η_1, y_2 to η_2; if $\partial\lambda/\partial\eta < 0$, then y decreases as η increases, and y_1 corresponds to η_2, y_2 to η_1. Both possibilities are covered by using the limits η_1, η_2 as given above and the modulus of $\partial\lambda/\partial\eta$.

In an exactly similar way, the integral on the right of (3) can be transformed by (b). We obtain

$$\iint_{\Delta} f\{\xi,\lambda(\xi,\eta)\}\left|\frac{\partial\lambda}{\partial\eta}\right| d\xi\,d\eta$$
$$= \iint_{D} f\{\theta(u,v),\phi(u,v)\}\left|\frac{\partial\lambda}{\partial\eta}\right|\left|\frac{\partial\theta}{\partial u}\right| du\,dv. \qquad (4)$$

Combining equations (3) and (4) and using (2), we obtain finally

$$\iint_{A} f(x,y)\,dx\,dy = \iint_{D} F(u,v)|J|\,du\,dv.$$

Consider the following example: by means of the transformation from x, y to u, v given by $y^2 = ux$, $x^2 = vy$, find the area of the region R bounded by the parabolas $y^2 = ax$, $y^2 = bx$, $x^2 = cy$, $x^2 = ey$, where $a > b > 0$, $c > e > 0$.

The area of R is given by $\iint_R dx\,dy$, taking $f(x,y)$ of the general theory equal to unity. Then (1) gives

$$\iint_{R} dx\,dy = \iint_{R'}\left|\frac{\partial(x,y)}{\partial(u,v)}\right| du\,dv, \qquad (5)$$

where R' is the region in the u–v plane corresponding to R in the x–y plane. First we sketch the regions R, R'. The region R is shaded in Fig. 12(i). The direct evaluation of $\iint_R dx\,dy$ is lengthy. If we integrate first with respect to y it is necessary to split the range of integration for x into three parts, with different limits for y in each case. The given transformation, however, produces a simple area R' over which we can integrate easily. Since $y^2 = ux$, the parabolas $y^2 = ax$, $y^2 = bx$ correspond to the lines $u = a$, $u = b$. Similarly the parabolas $x^2 = cy$, $x^2 = ey$ correspond by

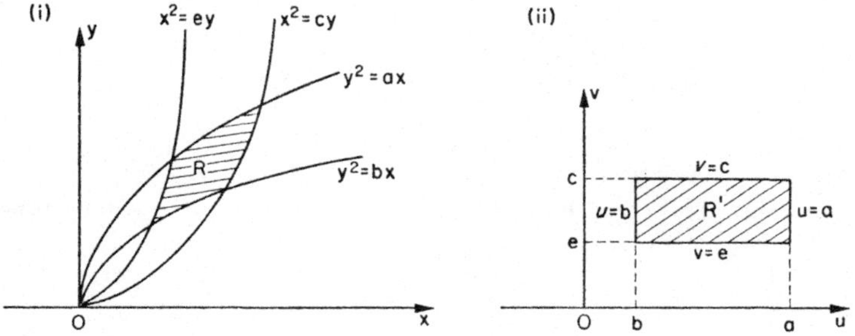

FIG. 12

$x^2 = vy$ to the lines $v = c$, $v = e$. The area R' is thus a rectangle (see Fig. 12(ii)). Hence, from (5)

$$\iint_R \mathrm{d}x\,\mathrm{d}y = \int_e^c \mathrm{d}v \int_b^a \left|\frac{\partial(x, y)}{\partial(u, v)}\right| \mathrm{d}u. \tag{6}$$

Note that the limits in (6) are constants. We can integrate with equal ease in either order; in (6) we have chosen to integrate first with respect to u.

The transformation is $u = y^2/x$, $v = x^2/y$. Using the property of Jacobians that $\partial(u,v)/\partial(x,y) = 1/\{\partial(x,y)/\partial(u,v)\}$, we find that

$$\frac{1}{J} = \frac{\partial(u, v)}{\partial(x, y)} = \left(\frac{-y^2}{x^2}\right)\left(\frac{-x^2}{y^2}\right) - \left(\frac{2y}{x}\right)\left(\frac{2x}{y}\right)$$

$$= -3.$$

So

$$J = \frac{\partial(x, y)}{\partial(u, v)} = -\frac{1}{3}.$$

Hence (6) becomes

$$\iint_R \mathrm{d}x\,\mathrm{d}y = \int_e^c \mathrm{d}v \int_b^a \frac{1}{3}\,\mathrm{d}u$$
$$= \tfrac{1}{3}(a-b)(c-e).$$

Exercises 4:4

1. Show in a diagram the region over which the integral

$$\int_0^a \mathrm{d}x \int_{3x}^{(4a-x)} [(x+y)/(3x-y+8a)^2]\,\mathrm{d}y$$

is taken. Prove that the substitution $x = u + v$, $y = 3v - u$ transforms the integral into

$$\int_0^a v\,\mathrm{d}v \int_{-v}^{0} [1/(u+2a)^2]\,\mathrm{d}u,$$

and hence evaluate the integral.

2. Show by using the transformation $y^4 = ux^3$, $x^3 = vy^2$, that the area in the first quadrant bounded by the curves $y^4 = ax^3$, $y^4 = bx^3$, $x^3 = ay^2$, $x^3 = by^2$, where $0 < a < b$, is $\frac{6}{35}(b^{5/6} - a^{5/6})(b^{7/6} - a^{7/6})$.
3. Using the transformation $u = x - y$, $v = x + 2y$, evaluate $\iint (x+y)^2\,\mathrm{d}x\,\mathrm{d}y$ over the region defined by $0 \leqq x - y \leqq 1$, $0 \leqq x + 2y \leqq 2$.
4. By means of the substitution $y + x = u$, $y - x = v$, evaluate the integral

$$\iint e^{(y-x)/(y+x)}\,\mathrm{d}x\,\mathrm{d}y,$$

taken over the triangle with vertices (0, 0), (0, 1), (1, 0).

4:5 The plane polar transformation

A special and very useful case of the general transformation given by 4:4(1) is the plane polar transformation $x = \rho\cos\psi$, $y = \rho\sin\psi$. Here the areas A and D are identical, and J is simply equal to ρ. For J is given by

$$J = \frac{\partial x}{\partial\rho}\frac{\partial y}{\partial\psi} - \frac{\partial x}{\partial\psi}\frac{\partial y}{\partial\rho} = \cos\psi\,.\,\rho\cos\psi - (-\rho\sin\psi)\sin\psi$$
$$= \rho(\cos^2\psi + \sin^2\psi)$$
$$= \rho.$$

So 4:4(1) becomes

$$\iint f(x, y)\,\mathrm{d}x\,\mathrm{d}y = \iint F(\rho, \psi)\rho\,\mathrm{d}\rho\,\mathrm{d}\psi, \tag{1}$$

where $F(\rho, \psi) = f[\rho \cos \psi, \rho \sin \psi]$. Equation (1) is easy to remember, since $\rho\,\mathrm{d}\rho\,\mathrm{d}\psi$ is the element of area in polar coordinates.

The plane polar transformation is particularly useful when integration is over a circle or part of a circle, and when the integrand contains functions of $x^2 + y^2$ or y/x. Normally, we integrate first with respect to ρ, and then with respect to ψ. The curves bounding the area of integration must be expressed in polar coordinates. We give two examples to illustrate the method.

EXAMPLE 1. *Evaluate* $\iint \log(x^2 + y^2)\,\mathrm{d}x\,\mathrm{d}y$ *over the area in the first quadrant between the circles* $x^2 + y^2 = a^2$, $x^2 + y^2 = b^2$, *where* $a > b$.

In polar coordinates, the circles have equations $\rho = a$, $\rho = b$. The area of integration A is shaded in Fig. 13. We measure ψ anti-clockwise from Ox so the limits for ψ are 0 and $\frac{1}{2}\pi$. For ψ

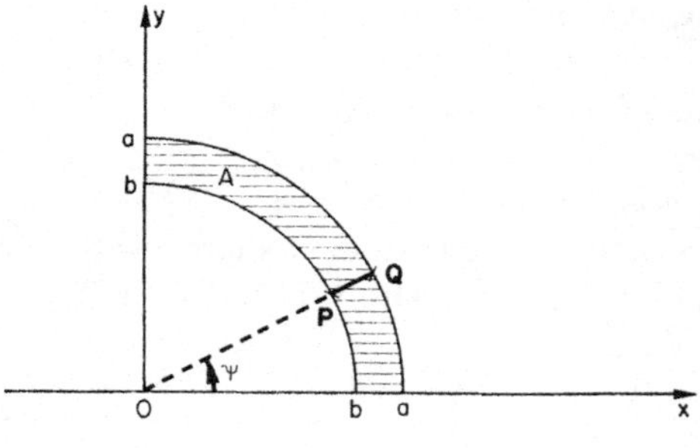

FIG. 13

held constant between 0 and $\frac{1}{2}\pi$, ρ varies from its value at P to its value at Q, that is from b to a. So

$$\iint_A \log(x^2 + y^2)\,\mathrm{d}x\,\mathrm{d}y = \iint_A \log(\rho^2)\rho\,\mathrm{d}\rho\,\mathrm{d}\psi$$

$$= \int_0^{\pi/2} \mathrm{d}\psi \int_b^a 2\rho \log \rho\,\mathrm{d}\rho.$$

Denoting the value of the integral on the left by I, and integrating on the right with respect to ρ by parts, we find

$$I = \int_0^{\pi/2} \left[\rho^2 \log \rho - \tfrac{1}{2}\rho^2\right]_b^a d\psi$$

$$= \left[a^2 \log a - \tfrac{1}{2}a^2 - b^2 \log b + \tfrac{1}{2}b^2\right] \int_0^{\pi/2} d\psi.$$

So $$I = \tfrac{1}{2}\pi \left[a^2 \log a - b^2 \log b - \tfrac{1}{2}(a^2 - b^2)\right].$$

EXAMPLE 2. *Show in a diagram the area of the x–y plane over which the double integral*

$$\int_0^a \int_y^{a+\sqrt{(a^2-y^2)}} [1/(4a^2 + x^2 + y^2)^2]\, dy\, dx$$

is taken. Evaluate this integral by changing to polar coordinates.

The first integration is with respect to x, since the first limits are functions of y. So for y held constant between 0 and a, x varies from $x = y$ to $x = a + \sqrt{(a^2 - y^2)}$; $x = y$ is a straight line through the origin and $x = a + \sqrt{(a^2 - y^2)}$ gives $(x - a)^2 = a^2 - y^2$, which is the equation of a circle with centre $(a, 0)$ and radius a. The area of integration is shaded in Fig. 14. Care must

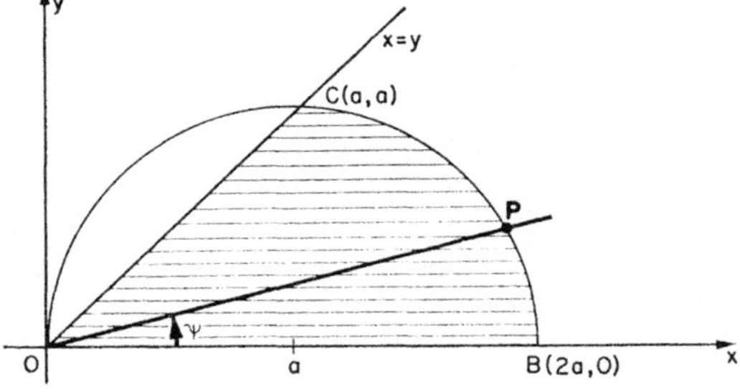

FIG. 14

be taken to obtain the correct area of integration here; at the upper limit for x, $x > a$, since $x = a + \sqrt{(a^2 - y^2)}$.

Changing to polar coordinates $x = \rho \cos \psi$, $y = \rho \sin \psi$, the line $x = y$ becomes $\psi = \frac{1}{4}\pi$. The circle $(x - a)^2 + y^2 = a^2$ becomes

$$(\rho \cos \psi - a)^2 + \rho^2 \sin^2 \psi = a^2,$$

so
$$\rho^2 - 2a\rho \cos \psi + a^2 = a^2,$$

and
$$\rho = 2a \cos \psi.$$

The limits for ψ are 0 and $\frac{1}{4}\pi$. To obtain these, imagine a line through the origin originally lying along OB, and then rotating anti-clockwise to the position OC: the general intermediate position is shown by OP in the figure. Then ψ changes from 0 at OB to $\frac{1}{4}\pi$ at OC. For ψ held constant between 0 and $\frac{1}{4}\pi$, ρ varies along the line OP from its value at O to its value at P, that is from 0 to $2a \cos \psi$. Applying the result in (1), we thus get

$$\int_0^a \int_y^{a+\sqrt{(a^2-y^2)}} \frac{dy\,dx}{(4a^2 + x^2 + y^2)^2}$$

$$= \int_0^{\pi/4} d\psi \int_0^{2a\cos\psi} \frac{\rho\, d\rho}{(4a^2 + \rho^2)^2}$$

$$= \int_0^{\pi/4} \left[-\frac{1}{2} \cdot \frac{1}{(4a^2 + \rho^2)} \right]_0^{2a\cos\psi} d\psi$$

$$= \frac{1}{8a^2} \int_0^{\pi/4} \left\{ 1 - \frac{1}{1 + \cos^2 \psi} \right\} d\psi$$

$$= \frac{1}{8a^2} \int_0^{\pi/4} d\psi - \frac{1}{8a^2} \int_0^{\pi/4} \frac{\sec^2 \psi}{1 + \sec^2 \psi} d\psi$$

$$= \frac{\pi}{32a^2} - \frac{1}{8a^2} \int_0^1 \frac{d(\tan \psi)}{2 + \tan^2 \psi}.$$

The integral on the right is now in standard form (as may easily be seen on writing $\tan\psi = u$, say); hence we obtain

$$\int_0^a\int_y^{a+\sqrt{(a^2-y^2)}}\frac{dy\,dx}{(4a^2+x^2+y^2)^2}$$
$$=\frac{\pi}{32a^2}-\frac{1}{8\sqrt{2}a^2}\left[\tan^{-1}\left(\frac{1}{\sqrt{2}}\tan\psi\right)\right]_{\tan\psi=0}^{1}$$
$$=\frac{1}{32a^2}\left[\pi-2\sqrt{2}\tan^{-1}\left(\frac{1}{\sqrt{2}}\right)\right].$$

We shall meet further examples of the polar transformation in § 4:8. It is essential that the reader be able to use this transformation with confidence and accuracy.

Exercises 4:5

1. Show that the value of the double integral $\iint\sqrt{(1-x^2-y^2)}\,dx\,dy$ taken over the interior of the circle $x^2+y^2=1$ is $\frac{2}{3}\pi$. [B.Sc.(Gen.)]
2. Express the integral
$$\int_0^1 dx\int_{\sqrt{3}x}^{\sqrt{(4-x^2)}}\{y^2/\sqrt{(x^2+y^2)}\}\,dy$$
in plane polar coordinates, and hence find its value.
3. The integral $\iint\{x^2/(x^2+y^2)\}\,dx\,dy$ is taken over the region in the first quadrant bounded by the ellipse $x^2/9a^2+y^2/a^2=1$, and the lines $x=3y$, $x=\sqrt{3}y$. By means of the transformation $x=3aX$, $y=aY$, convert the integral to one over a part of the circle $X^2+Y^2=1$. Hence, by using polar coordinates $X=\rho\cos\psi$, $Y=\rho\sin\psi$, find the value of the integral. [B.Sc.(Gen.)]
4. Indicate in a diagram the region of integration of the integral
$$\int_0^a xe^{-x^2}dx\int_{\sqrt{(ax-x^2)}}^{\sqrt{(a^2-x^2)}}\{ye^{-y^2}/(x^2+y^2)\}\,dy.$$
Change to polar coordinates, and hence show that the value of the integral is $[1-(1+a^2)\,e^{-a^2}]/4a^2$.

4:6 Green's Theorem

Green's Theorem provides an interesting and useful link between line integrals in two dimensions and double integrals.

We give a proof of the result, but readers wanting only applications of the theorem can omit this.

Suppose that $P(x,y)$, $Q(x,y)$ are single valued continuous functions of x and y defined throughout an area A and on its boundary curve C. Then Green's Theorem states that

$$\iint_A \left(\frac{\partial Q}{\partial x} - \frac{\partial P}{\partial y}\right) \mathrm{d}x\,\mathrm{d}y = \oint_C (P\,\mathrm{d}x + Q\,\mathrm{d}y). \tag{1}$$

To prove this, assume that C is not met by any line parallel to either axis in more than two points. (When this condition is not satisfied, the area A must be subdivided to give regions whose boundaries do satisfy it; the separate double and line integrals are then added together; see Fig. 15.) We show first that

$$\iint_A \frac{\partial Q}{\partial x} \mathrm{d}x\,\mathrm{d}y = \oint_C Q\,\mathrm{d}y.$$

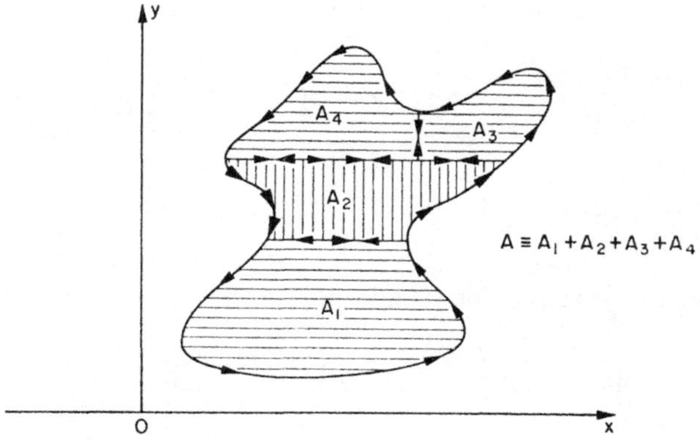

FIG. 15

In Fig. 16, the extreme values of x on C are a, b, and the extreme values of y are c, e. The curve C may be divided into two parts FBE, FDE. Suppose that on FBE, x is given by $x = \theta_1(y)$, and that

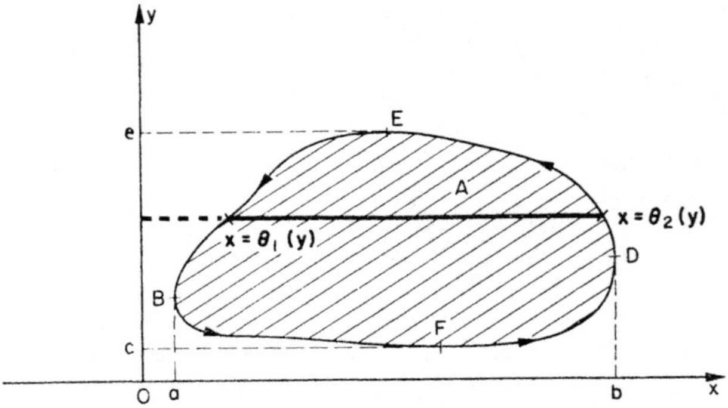

FIG. 16

on FDE, x is given by $x = \theta_2(y)$. Then

$$\iint_A \frac{\partial Q}{\partial x} \mathrm{d}x\, \mathrm{d}y = \int_c^e \mathrm{d}y \int_{x=\theta_1(y)}^{x=\theta_2(y)} \frac{\partial Q}{\partial x} \mathrm{d}x$$

$$= \int_c^e \Big[Q(x, y) \Big]_{x=\theta_1(y)}^{x=\theta_2(y)} \mathrm{d}y$$

$$= \int_c^e [Q\{\theta_2(y), y\} - Q\{\theta_1(y), y\}]\, \mathrm{d}y$$

or $$\iint_A \frac{\partial Q}{\partial x} \mathrm{d}x\, \mathrm{d}y = \int_c^e [Q_2(y) - Q_1(y)]\, \mathrm{d}y, \text{ say.} \qquad (2)$$

Turning now to the line integral,

$$\oint_C Q\, \mathrm{d}y = \int_{FDE} Q(x, y)\, \mathrm{d}y + \int_{EBF} Q(x, y)\, \mathrm{d}y$$

$$= \int_c^e Q[\theta_2(y), y]\, \mathrm{d}y + \int_e^c Q[\theta_1(y), y]\, \mathrm{d}y$$

$$= \int_c^e \{Q_2(y) - Q_1(y)\}\, \mathrm{d}y. \qquad (3)$$

Comparing (2) and (3),

$$\oint_C Q\,\mathrm{d}y = \iint_A \frac{\partial Q}{\partial x}\mathrm{d}x\,\mathrm{d}y. \tag{4}$$

Similarly, if we write $y = \phi_1(x)$ on BFD and $y = \phi_2(x)$ on BED, we find that

$$\iint_A \frac{\partial P}{\partial y}\mathrm{d}x\,\mathrm{d}y = \int_a^b \mathrm{d}x \int_{y=\phi_1(x)}^{y=\phi_2(x)} \frac{\partial P}{\partial y}\mathrm{d}y$$

$$= \int_a^b [P\{x_1, \phi_2(x)\} - P\{x_1, \phi_1(x)\}]\,\mathrm{d}x$$

$$= \int_a^b \{P_2(x) - P_1(x)\}\,\mathrm{d}x$$

$$= \int_{BED} P\,\mathrm{d}x - \int_{BFD} P\,\mathrm{d}x.$$

Therefore

$$\iint_A \frac{\partial P}{\partial y}\mathrm{d}x\,\mathrm{d}y = -\oint_C P\,\mathrm{d}x. \tag{5}$$

Hence, combining (4) and (5), we obtain Green's Theorem

$$\oint_C [P(x, y)\,\mathrm{d}x + Q(x, y)\,\mathrm{d}y] = \iint_A \left(\frac{\partial Q}{\partial x} - \frac{\partial P}{\partial y}\right)\mathrm{d}x\,\mathrm{d}y. \tag{6}$$

It is often shorter to evaluate a double integral rather than a line integral. For example, suppose we require

$$\oint_C [(2y^2 - x^2)\,\mathrm{d}x + (3x^2 - y^2)\,\mathrm{d}y],$$

where C is the circle $(x - a)^2 + y^2 = a^2$. The direct method is to put $x = a(1 + \cos\theta)$, $y = a \sin\theta$, the limits for θ in the line integral being 0 and 2π (see Fig. 17). A shorter method is given

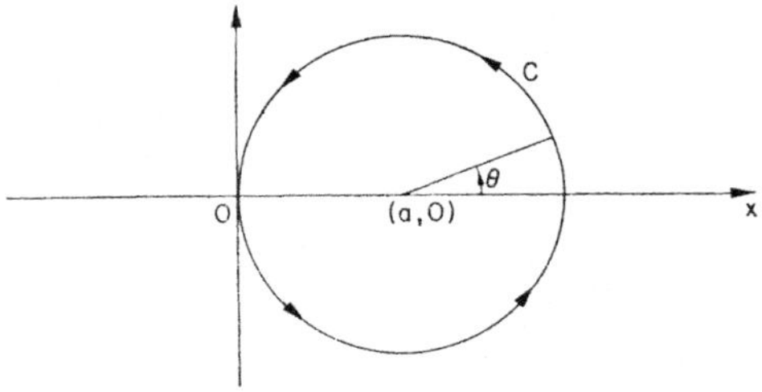

FIG. 17

by using Green's Theorem. Comparing with (6), we have $P = 2y^2 - x^2$, $Q = 3x^2 - y^2$. So (6) gives

$$\oint_C [(2y^2 - x^2)\,dx + (3x^2 - y^2)\,dy] = \iint_A (6x - 4y)\,dx\,dy,$$

where A is the area enclosed by C.

From symmetry, we can say at once that $\iint_A y\,dx\,dy = 0$. To evaluate $\iint_A 6x\,dx\,dy$ we can integrate in cartesians, or, better, change to polar coordinates, obtaining the value $6\pi a^3$. So the given line integral has the value $6\pi a^3$.

Exercises 4:6

1. Calculate directly $\oint_C[(x^3 + y^2)\,dx + (x^2 + y^3)\,dy]$, where C is the boundary of the pentagon whose vertices are the points (0, 0), (1, 0), (2, 1), (1, 2), (0, 1). Verify the result by applying Green's Theorem.
2. Use Green's Theorem to evaluate the line integral
$$\oint_C[(x^3 + 6x^2y + y^2 - 3y)\,dx + (2x^3 - 4xy^2 - 2xy + 3y)\,dy],$$
where C is the circle $(x - 2)^2 + (y - 2)^2 = 4$.
3. Use Green's Theorem to evaluate $\oint_C[(y - \cos x)\,dx + \sin x\,dy]$, where C is the perimeter of the triangle formed by the lines $y = 2x/\pi$, $x = \frac{1}{2}\pi$, $y = 0$. [B.Sc.(Eng.)]

4:7 Triple integrals

Double integrals are merely a special case of multiple integrals. The general multiple integral involves a function of n variables, and so n integrations have to be performed. We consider now the triple integral, when $n = 3$. In cartesian coordinates a triple integral is of the form

$$\iiint g(x, y, z)\,\mathrm{d}x\,\mathrm{d}y\,\mathrm{d}z.$$

Whereas the double integral is evaluated over an area in two dimensions, the triple integral is taken through a volume in three dimensions.

Triple integrals have many applications in physical problems. Perhaps the simplest example is the use of a triple integral to obtain the mass of a given volume of variable density. Suppose that the volume V of § 4:1, Fig. 7, is occupied by matter whose density at the point (x, y, z) is $m(x, y, z)$. Then the mass of the element of volume $\mathrm{d}x\,\mathrm{d}y\,\mathrm{d}z$ is $m(x, y, z)\mathrm{d}x\,\mathrm{d}y\,\mathrm{d}z$, and the total mass M is given by

$$M = \iiint_V m(x, y, z)\,\mathrm{d}x\,\mathrm{d}y\,\mathrm{d}z.$$

Let us integrate first with respect to z; this involves holding both x and y constant, that is, fixing a point in A. Then z varies from its value 0 on the x–y plane to its value $f(x, y)$ on the surface A'. So

$$M = \iiint_V m(x, y, z)\,\mathrm{d}x\,\mathrm{d}y\,\mathrm{d}z = \iint_A \mathrm{d}x\,\mathrm{d}y \int_0^{f(x,y)} m\,\mathrm{d}z$$

$$= \iint_A h(x, y)\,\mathrm{d}x\,\mathrm{d}y,$$

where $$h(x, y) = \int_0^{f(x,y)} m(x, y, z)\,\mathrm{d}z.$$

Thus one integration reduces the triple integral to the form of double integral we have already studied. In general, if we integrate in the order z, y, x, then the limits for z will be functions of *both* x and y, those for y will be functions of x, and those for x will be constants. As an example, we show how to evaluate the integral $\iiint(x + y + z + 1)^{-3}\mathrm{d}z\mathrm{d}y\mathrm{d}x$ taken through the volume enclosed by the planes $x = 0$, $y = 0$, $z = 0$, $x + y + z = 1$.

The volume of integration is the regular tetrahedron $OABC$ shown in Fig. 18, bounded by the coordinate planes and the

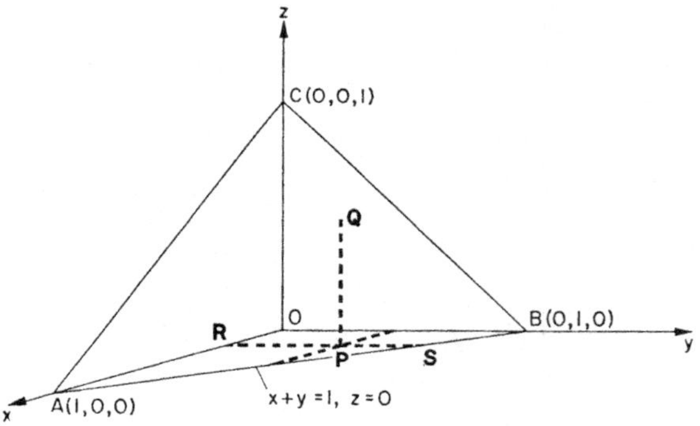

FIG. 18

plane ABC, which has equation $x + y + z = 1$. For x and y held constant in triangle OAB, at P say, z varies from its value at P to its value at Q, where PQ is parallel to Oz. So z varies from 0 (at P) to $1 - x - y$ (at Q). Now for x held constant on OA, at R say, where RP is parallel to Oy, y varies from its value at R to its value at S where RP meets AB; hence y varies from 0 (at R) to $1 - x$ (at S). Finally, x varies along OA from $x = 0$ at O to $x = 1$ at A. So

$$\iiint(x + y + z + 1)^{-3}\,\mathrm{d}z\,\mathrm{d}y\,\mathrm{d}x$$

$$= \int_0^1 \mathrm{d}x \int_0^{1-x} \mathrm{d}y \int_0^{1-x-y} (x + y + z + 1)^{-3}\,\mathrm{d}z.$$

Denoting the value of the integral by I, we now have

$$I = \int_0^1 \mathrm{d}x \int_0^{1-x} \left[-\tfrac{1}{2}(x+y+z+1)^{-2} \right]_{z=0}^{z=1-x-y} \mathrm{d}y$$

$$= \frac{1}{2}\int_0^1 \mathrm{d}x \int_0^{1-x} \left[\frac{1}{(x+y+1)^2} - \frac{1}{4} \right] \mathrm{d}y$$

$$= \frac{1}{2}\int_0^1 \left[-\frac{1}{(x+y+1)} - \frac{y}{4} \right]_{y=0}^{y=1-x} \mathrm{d}x$$

$$= \frac{1}{2}\int_0^1 \left[\frac{1}{1+x} - \frac{1}{2} - \frac{(1-x)}{4} \right] \mathrm{d}x$$

$$= \frac{1}{2}\left[\log(1+x) - \frac{3x}{4} + \frac{x^2}{8} \right]_0^1.$$

Therefore $$I = \tfrac{1}{2}(\log 2 - \tfrac{5}{8}).$$

For triple integrals, the result corresponding to 4:4(1) is

$$\iiint_V (x, y, z)\,\mathrm{d}x\,\mathrm{d}y\,\mathrm{d}z = \iiint_{V'} F(u, v, w)|J|\,\mathrm{d}u\,\mathrm{d}v\,\mathrm{d}w, \qquad (1)$$

where x, y, z are each functions of u, v, and w, V' is the volume in the u, v, w system corresponding to V, and J is the Jacobian

$$\frac{\partial(x,\ y,\ z)}{\partial(u,\ v,\ w)}.$$

The very useful spherical polar transformation deserves attention. We transform from x, y, z to r, θ, ψ, where

$$x = r\sin\theta\cos\psi, \quad y = r\sin\theta\sin\psi, \quad z = r\cos\theta.$$

In Fig. 19, where PN is the perpendicular from $P(x, y, z)$ to the plane xOy, note carefully the directions in which θ, ψ are

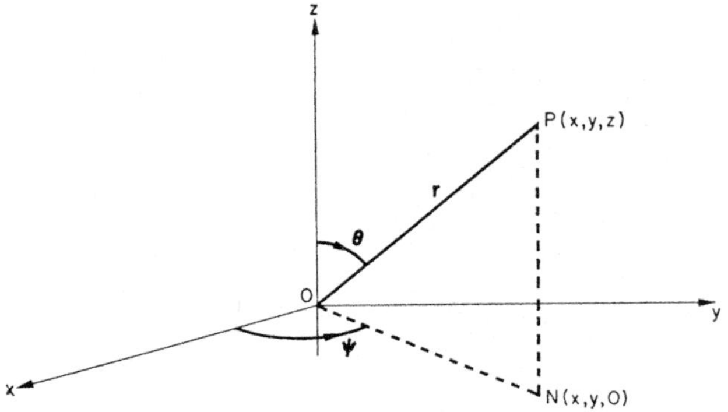

FIG. 19

measured. Evaluating the Jacobian J, we find at once

$$J = \begin{vmatrix} \sin\theta\cos\psi, & r\cos\theta\cos\psi, & -r\sin\theta\sin\psi \\ \sin\theta\sin\psi, & r\cos\theta\sin\psi, & r\sin\theta\cos\psi \\ \cos\theta, & -r\sin\theta, & 0 \end{vmatrix}$$

$$= r^2\sin\theta.$$

So (1) becomes

$$\iiint_V (x, y, z)\,\mathrm{d}x\,\mathrm{d}y\,\mathrm{d}z = \iiint_V F(r, \theta, \psi)r^2|\sin\theta|\,\mathrm{d}r\,\mathrm{d}\theta\,\mathrm{d}\psi. \quad (2)$$

In fact, we can replace $|\sin\theta|$ in (2) by $\sin\theta$, since all points in V are reached by taking θ between 0 and π, and allowing ψ to vary between 0 and 2π. Also, in this transformation V and V' are identical; $r^2|\sin\theta|\mathrm{d}r\,\mathrm{d}\theta\,\mathrm{d}\psi$ is, of course, the element of volume in spherical polar coordinates.

As an example, we evaluate $\iiint xyz\,\mathrm{d}x\,\mathrm{d}y\mathrm{d}z$ taken through the positive octant of the unit sphere $x^2 + y^2 + z^2 = 1$. In spherical polars, the sphere is given by $r = 1$, and the positive octant of it implies that the limits for both θ and ψ are 0 and $\frac{1}{2}\pi$. So

$$\iiint_V xyz\,dx\,dy\,dz$$

$$= \iiint_V r^3 \sin^2\theta \cos\psi \sin\psi \cos\theta\, r^2 \sin\theta\, dr\, d\theta\, d\psi$$

$$= \int_0^{\pi/2} \sin^3\theta \cos\theta\, d\theta \int_0^{\pi/2} \cos\psi \sin\psi\, d\psi \int_0^1 r^5 dr.$$

Since the limits for all three integrals are constants, the triple integral reduces at once to the product of three single integrals. Therefore,

$$\iiint_V xyz\,dx\,dy\,dz = \left[\frac{\sin^4\theta}{4}\right]_0^{\pi/2} \cdot \left[\frac{\sin^2\psi}{2}\right]_0^{\pi/2} \cdot \left[\frac{r^6}{6}\right]_0^1$$

$$= 1/48.$$

When the region of integration is a sphere or part of a sphere, the spherical polar transformation provides the simplest means of evaluation in most cases.

Exercises 4:7

1. Evaluate $\iiint xy^2\, dx\, dy\, dz$ through the interior of the pyramid bounded by the planes $x = a$, $y = 0$, $y = x$, $z = 0$, $z = x$.
2. Evaluate $\iiint z^2\, dx\, dy\, dz$ through the volume enclosed by the planes $y = 0$, $x = 0$, $z = 1$, $2x + y - 2z = 0$.
3. Evaluate $\iiint yz\, dx\, dy\, dz$ taken throughout the prism with sides parallel to the z-axis, whose base is the triangle with vertices at (0, 0, 0), (1, 0, 0), (0, 1, 0), and whose top is the triangle with vertices (0, 0, 2), (1, 0, 1), (0, 1, 1). [B.Sc.(Eng.)]
4. The integral $\iiint (2x + 3y + 4z)^2\, dx\, dy\, dz$ is taken through the volume enclosed by the ellipsoid $x^2/a^2 + y^2/b^2 + z^2/c^2 = 1$. By means of the transformation $x = aX$, $y = bY$, $z = cZ$, convert this integral to one taken through the unit sphere. Using spherical polar coordinates, show that the value of this integral is $\frac{4}{15}\pi abc(4a^2 + 9b^2 + 16c^2)$.

4:8 Types of problem

We consider now three of the most common types of problem involving double and triple integrals.

(i) The problem may be simply the evaluation of a given double or triple integral; this may involve a change of the order of integration or the use of a suitable change of variables. In any case, the first step is to make a sketch of the area or volume of integration. The reader is referred to the worked examples in §§ 4:2, 4:3, 4:4, 4:5 which are typical of these problems.

(ii) The problem may be that of finding the volume of a solid, which may be specified in physical or analytical terms. Mathematically, of course, the volume V of a region R of three-dimensional space is *defined* by the *triple integral*

$$V = \iiint_R \mathrm{d}x\,\mathrm{d}y\,\mathrm{d}z.$$

However, when the bounding surfaces of the region R are precisely given, either physically or analytically, it is frequently possible to write down *immediately* a *double integral* expression for V of the form $\iint_A f(x,y)\,\mathrm{d}x\,\mathrm{d}y$, where A is an area in the x–y plane: compare the reduction of a triple integral to a double integral at the first integration, discussed on p. 73. We illustrate this method by two examples.

EXAMPLE 1. *A cylindrical hole of diameter a is drilled through a uniform solid sphere of radius a, the centre of the sphere lying on the surface of the cylinder. Find the volume of material removed from the sphere.* [B.Sc.(Gen.)]

We have first to express the problem analytically. The equation of a sphere, centre the origin, radius a, is $x^2 + y^2 + z^2 = a^2$. This is the simplest form we can choose for the sphere. We can take the cylinder with generators parallel to the z-axis, and with its cross-section in the x–y plane having equation

$$x^2 + (y - \tfrac{1}{2}a)^2 = (\tfrac{1}{2}a)^2. \qquad (1)$$

In Fig. 20, O is the centre of the sphere, O' the centre $(0, \tfrac{1}{2}a, 0)$ of the circle (1) in the x–y plane. The volume we want is, by symmetry, four times the volume common to the cylinder and the sphere in the positive octant. We shall thus integrate over the

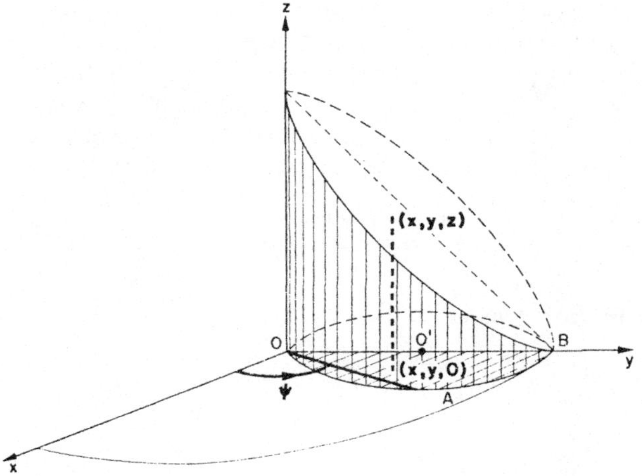

FIG. 20

shaded area OAB in the figure. So, if V is the volume required,

$$V = 4\iint_{OAB} z\,\mathrm{d}x\,\mathrm{d}y,$$

where z is the value of z on the sphere corresponding to the point (x, y) of the shaded area OAB. From the equation of the sphere, the value of z on the surface in the positive octant is

$$z = +\sqrt{(a^2 - x^2 - y^2)}.$$

So
$$V = 4\iint_{OAB} \sqrt{(a^2 - x^2 - y^2)}\,\mathrm{d}x\,\mathrm{d}y. \quad (2)$$

If we integrate first with respect to x, then on inserting limits we find

$$V = 4\int_0^a \mathrm{d}y \int_0^{\sqrt{(ay-y^2)}} \sqrt{(a^2 - x^2 - y^2)}\,\mathrm{d}x, \quad (3)$$

the upper limit for x being given by (1) above, which reduces to

$$x^2 + y^2 - ay = 0. \quad (4)$$

This is a case where integration in cartesian coordinates is unnecessarily complicated; instead of evaluating (3) as it stands we transform to polar coordinates. The equation of the circle (4) reduces to $\rho = a\sin\psi$, when we put $x = \rho\cos\psi$, $y = \rho\sin\psi$. The limits of integration for ψ are 0 and $\frac{1}{2}\pi$; for ψ held constant between 0 and $\frac{1}{2}\pi$, ρ varies from 0 to $a\sin\psi$. So

$$V = 4\int_0^{\pi/2} d\psi \int_0^{a\sin\psi} \sqrt{(a^2-\rho^2)}\rho\,d\rho,$$

using the result given in 4:5(1). Thus

$$V = 4\int_0^{\pi/2}\left[-\tfrac{1}{3}(a^2-\rho^2)^{3/2}\right]_0^{a\sin\psi} d\psi$$

$$= \frac{4a^3}{3}\int_0^{\pi/2}(1-\cos^3\psi)\,d\psi$$

$$= \frac{4a^3}{3}\left(\frac{\pi}{2}-\frac{2}{3}\right)$$

$$= \frac{2a^3}{9}(3\pi-4).$$

EXAMPLE 2. *A cylinder is formed with generators parallel to the z-axis and passing through the curve of intersection of the surface* $x^2 + y^2 = 4az$ *and the plane* $x + y + z = 2a$. *Show that the finite volume lying in the positive octant and bounded by the coordinate planes, the given plane and the cylinder may be expressed as a double integral in the form*

$$2\int_{\pi/6}^{\pi/4}\int_{2a\,\mathrm{cosec}\,\psi}^{4a}[6a-\rho(\cos\psi+\sin\psi)]\,\rho\,d\rho\,d\psi,$$

where ρ, ψ *are plane polar coordinates in the x–y plane with the point* $(-2a, -2a, 0)$ *as origin and the line* $y + 2a = z = 0$ *as initial line. Evaluate the integral.* [B.Sc.(Eng.)]

The surface $x^2 + y^2 = 4az$ meets the plane $x + y + z = 2a$ where

$$x^2 + y^2 = 4a(2a - x - y).$$

This gives, on rearrangement,

$$(x + 2a)^2 + (y + 2a)^2 = 16a^2. \tag{4a}$$

Thus the equation of the cylinder with generators parallel to O is given by (4a). Its cross-section by the plane $z = 0$ is a circle, centre $(-2a, -2a, 0)$, and radius $4a$. In Fig. 21, C is the point

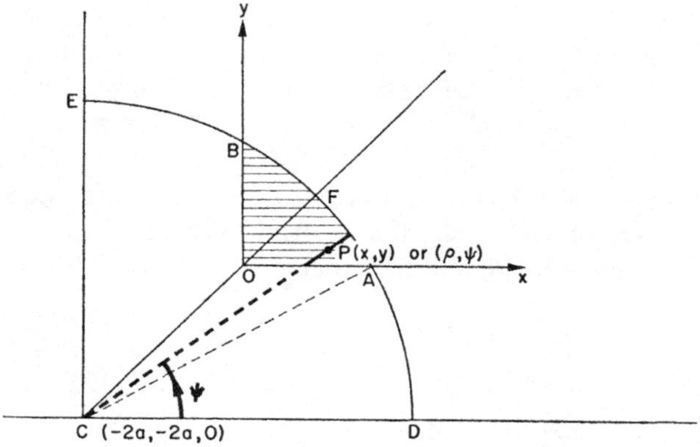

FIG. 21

$(-2a, -2a)$ in the x–y plane, and OAB is the area of integration. If V is the volume required, then

$$V = \iint_{OAB} z\,dx\,dy,$$

where z takes its value on the plane $x + y + z = 2a$. Therefore

$$V = \iint_{OAB} (2a - x - y)\,dx\,dy. \tag{5}$$

We now change to the polar coordinates suggested in the question;

thus we take C as origin and CD as initial line. If P is the point (ρ, ψ) in polar coordinates and (x, y) in cartesians, we see that

$$\rho \cos \psi = 2a + x, \quad \rho \sin \psi = 2a + y.$$

The circle (4a) transforms to $\rho = 4a$, and (5) becomes

$$V = \iint_{OAB} [2a - (\rho \cos \psi - 2a) - (\rho \sin \psi - 2a)]\rho \, d\rho \, d\psi,$$

i.e.
$$V = \iint_{OAB} [6a - \rho(\cos \psi + \sin \psi)]\rho \, d\rho \, d\psi. \tag{6}$$

From the form of (6) we see that there is symmetry about $\psi = \frac{1}{4}\pi$, that is about the line CF in the figure. So V is given by twice the integral taken over OAF. On CA, $\psi = \frac{1}{6}\pi$ (for A is distant $2a$ from CD and $CA = 4a$); thus the limits for ψ are $\frac{1}{6}\pi$ and $\frac{1}{4}\pi$. For ψ held constant between $\frac{1}{6}\pi$ and $\frac{1}{4}\pi$, ρ varies from its value on OA to its value on the circle, that is from $2a \operatorname{cosec} \psi$ to $4a$. Hence, inserting limits in (6), we find that

$$V = 2\int_{\pi/6}^{\pi/4}\int_{2a \operatorname{cosec} \psi}^{4a} [6a - \rho(\cos \psi + \sin \psi)]\rho \, d\rho \, d\psi$$

$$= 2\int_{\pi/6}^{\pi/4}\left[3a\rho^2 - \frac{\rho^3}{3}\left(\cos \psi + \sin \psi\right)\right]_{2a \operatorname{cosec} \psi}^{4a} d\psi$$

$$= 2a^3\int_{\pi/6}^{\pi/4}\left[48 - 12 \operatorname{cosec}^2 \psi - \right.$$

$$\left. - \frac{1}{3}\left(\cos \psi + \sin \psi\right)\left(64 - 8 \operatorname{cosec}^2 \psi\right)\right] d\psi$$

$$= 2a^3\left[48\psi + 12 \cot \psi - \frac{64}{3}\left(\sin \psi - \cos \psi\right) - \right.$$

$$\left. - \frac{8}{3}\cot \psi - \frac{4}{3}\cot^2 \psi\right]_{\pi/6}^{\pi/4}.$$

Inserting the upper and lower limits and simplifying, we find that

$$V = \tfrac{2}{3}a^3(68 + 12\pi - 60\sqrt{3}).$$

(iii) Multiple integrals are of great use in many types of physical problem. We note especially here the value of double and triple integrals in finding centres of gravity and moments of inertia.

Suppose that we have a solid occupying a volume V, defined in terms of cartesian coordinates x, y, z. We shall take the density at the point (x, y, z) of the solid as $m(x,y,z)$, so that in general m varies from point to point of V. If the centre of gravity of the solid is at the point $(\bar{x}, \bar{y}, \bar{z})$, we have the equations

$$M = \iiint_V m \, \mathrm{d}x \, \mathrm{d}y \, \mathrm{d}z, \tag{7}$$

$$\left.\begin{aligned} M\bar{x} &= \iiint_V mx \, \mathrm{d}x \, \mathrm{d}y \, \mathrm{d}z, \\ M\bar{y} &= \iiint_V my \, \mathrm{d}x \, \mathrm{d}y \, \mathrm{d}z, \\ M\bar{z} &= \iiint_V mz \, \mathrm{d}x \, \mathrm{d}y \, \mathrm{d}z, \end{aligned}\right\} \tag{8}$$

where M is the mass of the solid.

The integral $\iiint_V x \, \mathrm{d}x \, \mathrm{d}y \, \mathrm{d}z$ gives the *first moment* of the volume V about the plane $x = 0$. If m is constant, then $M = Vm$, and (7), (8) can be expressed in terms of V instead of M.

For a plane mass distribution, in the plane $z = 0$, of surface density σ and area A, we have

$$\left.\begin{aligned} M &= \iint_A \sigma \, \mathrm{d}x \, \mathrm{d}y, \\ M\bar{x} &= \iint_A \sigma x \, \mathrm{d}x \, \mathrm{d}y, \\ M\bar{y} &= \iint_A \sigma y \, \mathrm{d}x \, \mathrm{d}y. \end{aligned}\right\} \tag{9}$$

The first moment of the plane mass distribution about the y-axis is $\iint_A \sigma x \, dx \, dy$, and about the x-axis it is $\iint_A \sigma y \, dx \, dy$.

Turning now to *second moments*, let I_x, I_y, I_z be the moments of inertia of a solid of volume V about the x-, y-, and z-axes respectively. Then the equations giving I_x, I_y, I_z are

$$I_x = \iiint_V m(y^2 + z^2) \, dx \, dy \, dz, \tag{10a}$$

$$I_y = \iiint_V m(z^2 + x^2) \, dx \, dy \, dz, \tag{10b}$$

$$I_z = \iiint_V m(x^2 + y^2) \, dx \, dy \, dz; \tag{10c}$$

for, in (10a), the element of mass is $m \, dx \, dy \, dz$ and the square of its distance from the x-axis is $(y^2 + z^2)$. It is usual to introduce the radius of gyration, and write $I_x = Mk_x{}^2$, where k_x is the radius of gyration of the solid about the x-axis. The quantities k_y, k_z are defined similarly.

The corresponding equations for a plane mass distribution of area A and surface density σ in the plane $z = 0$ are

$$I_x = \iint_A \sigma y^2 \, dx \, dy, \tag{11a}$$

$$I_y = \iint_A \sigma x^2 \, dx \, dy, \tag{11b}$$

$$I_z = \iint_A \sigma(x^2 + y^2) \, dx \, dy. \tag{11c}$$

In the evaluation of these integrals it may, of course, be necessary, or simpler, in a particular case, to employ a change of variables before integration.

EXAMPLE 1. *The density at any point P of a solid bounded by the surface* $x^2/a^2 + y^2/b^2 + z^2/c^2 = 1$ *is proportional to the product of the distances of P from the three coordinate planes. Show that the moment of inertia of the solid about the x-axis is* $\frac{1}{4}M(b^2 + c^2)$, *where M is the mass of the solid.*

The given solid is bounded by the ellipsoid $x^2/a^2 + y^2/b^2 + z^2/c^2 = 1$. Denote the volume of the solid by V, and write $m = kxyz$, where k is a constant. Then we have

$$M = \iiint_V kxyz\,dx\,dy\,dz, \tag{12}$$

$$I_x = \iiint_V kxyz(y^2 + z^2)\,dx\,dy\,dz. \tag{13}$$

Equations (12) and (13) are to be taken through the volume V. If we integrate first with respect to z, we then have a double integral over the ellipse $x^2/a^2 + y^2/b^2 = 1$ to deal with. It is far simpler here to change the variables by means of the transformation $x = aX$, $y = bY$, $z = cZ$. This transforms the ellipsoid into the unit sphere $X^2 + Y^2 + Z^2 = 1$. Denoting the volume of this sphere by V',

$$M = ka^2b^2c^2 \iiint_{V'} XYZ\,dX\,dY\,dZ,$$

$$I_x = ka^2b^2c^2 \iiint_{V'} XYZ(b^2Y^2 + c^2Z^2)\,dX\,dY\,dZ.$$

Now we change variables again, this time to spherical polar coordinates. Then

$$M = ka^2b^2c^2 \iiint_{V'} r^5 \sin^3\theta \cos\theta \sin\psi \cos\psi\,dr\,d\theta\,d\psi, \tag{14}$$

$$I_x = ka^2b^2c^2 \iiint_{V'} r^7 \sin^3\theta \cos\theta \sin\psi \cos\psi\,(b^2 \sin^2\theta \sin^2\psi + c^2\cos^2\theta)\,dr\,d\theta\,d\psi. \tag{15}$$

In each case, we can replace the integral by eight times the integral through the positive octant. Equation (14) thus becomes

$$M = 8ka^2b^2c^2\int_0^{\pi/2}\sin^3\theta\cos\theta\,d\theta\int_0^{\pi/2}\sin\psi\cos\psi\,d\psi\int_0^1 r^5\,dr$$

$$= 8ka^2b^2c^2\left[\frac{\sin^4\theta}{4}\right]_0^{\pi/2}\cdot\left[\frac{\sin^2\psi}{2}\right]_0^{\pi/2}\cdot\left[\frac{r^6}{6}\right]_0^1.$$

Thus
$$M = \tfrac{1}{6}ka^2b^2c^2.$$

In the same way (15) becomes

$$I_x = 8ka^2b^2c^2\int_0^{\pi/2}b^2\sin^5\theta\cos\theta\,d\theta\int_0^{\pi/2}\sin^3\psi\cos\psi\,d\psi\int_0^1 r^7\,dr +$$

$$+ 8ka^2b^2c^2\int_0^{\pi/2}c^2\sin^3\theta\cos^3\theta\,d\theta\int_0^{\pi/2}\sin\psi\cos\psi\,d\psi\int_0^1 r^7\,dr$$

$$= 8ka^2b^2c^2\left\{b^2\left[\frac{\sin^6\theta}{6}\right]_0^{\pi/2}\left[\frac{\sin^4\psi}{4}\right]_0^{\pi/2}\left[\frac{r^8}{8}\right]_0^1 + \right.$$

$$\left. + c^2\cdot\frac{2}{6}\left[\frac{\sin^4\theta}{4}\right]_0^{\pi/2}\left[\frac{\sin^2\psi}{2}\right]_0^{\pi/2}\left[\frac{r^8}{8}\right]_0^1\right\}$$

$$= ka^2b^2c^2\left[b^2\cdot\frac{1}{24} + c^2\cdot\frac{1}{24}\right].$$

Therefore
$$I_x = \frac{1}{24}ka^2b^2c^2(b^2 + c^2).$$

Hence, using the value of M just obtained, we find that

$$I_x = \tfrac{1}{4}M(b^2 + c^2).$$

EXAMPLE 2. *Find the second moment of the area between the loop of the curve $\rho = a\cos 2\psi$ for which $-\frac{1}{4}\pi \leqq \psi \leqq \frac{1}{4}\pi$, and the circle $\rho = 2a\cos\psi$ about a line through the origin perpendicular to the plane of the area.* [B.Sc.(Eng.)]

The area required is shaded in Fig. 22. Denote the shaded area by A. Taking the plane of the area as the (x, y) plane, so that ρ, ψ

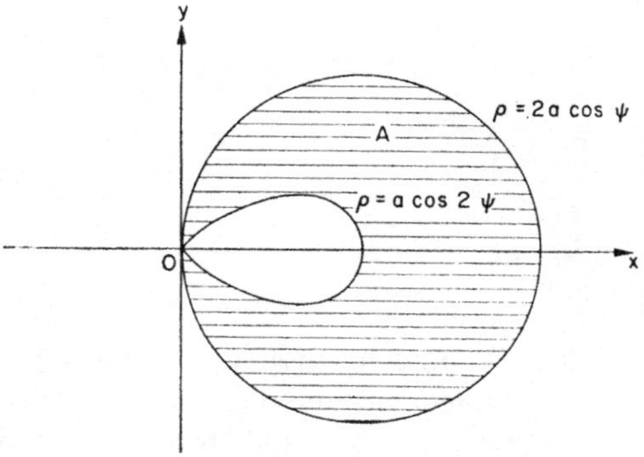

FIG. 22

are the usual polar coordinates given by $x = \rho\cos\psi$, $y = \rho\sin\psi$, then the line about which the second moment is required is the z-axis. Hence, using equation (11c),

$$I_z = \iint_A (x^2 + y^2)\,\mathrm{d}x\,\mathrm{d}y$$

$$= \iint_A \rho^2 \,.\, \rho\,\mathrm{d}\rho\,\mathrm{d}\psi.$$

If we now denote the area of the circle by A_1 and the area of the loop of the curve $\rho = a\cos 2\psi$ by A_2, then $A = A_1 - A_2$. So

$$I_z = \iint_{A_1} \rho^3\,\mathrm{d}\rho\,\mathrm{d}\psi - \iint_{A_2} \rho^3\,\mathrm{d}\rho\,\mathrm{d}\psi.$$

Using the symmetry of the area A, we insert the limits of integration, obtaining

$$I_z = 2\int_0^{\pi/2} \mathrm{d}\psi \int_0^{2a\cos\psi} \rho^3\,\mathrm{d}\rho - 2\int_0^{\pi/4} \mathrm{d}\psi \int_a^{a\cos 2\psi} \rho^3\,\mathrm{d}\rho$$

$$= 2\int_0^{\pi/2} \frac{1}{4}\Big[\rho^4\Big]_0^{2a\cos\psi} \mathrm{d}\psi - 2\int_0^{\pi/4} \frac{1}{4}\Big[\rho^4\Big]_0^{a\cos 2\psi} \mathrm{d}\psi$$

$$= 8a^4 \int_0^{\pi/2} \cos^4\psi\,\mathrm{d}\psi - \frac{1}{2}a^4 \int_0^{\pi/4} \cos^4 2\psi\,\mathrm{d}\psi$$

$$= 8a^4 \cdot \frac{3}{4} \cdot \frac{1}{2} \cdot \frac{\pi}{2} - \frac{1}{2}a^4 \int_0^{\pi/2} \cos^4 \phi \cdot \frac{1}{2} d\phi \quad \text{(putting } 2\psi = \phi\text{)}$$

$$= \frac{3\pi a^4}{2} - \frac{1}{4}a^4 \frac{3}{4} \cdot \frac{1}{2} \cdot \frac{\pi}{2}$$

$$= 3\pi a^4 \left(\frac{1}{2} - \frac{1}{64}\right).$$

Therefore $\qquad I_z = \frac{93}{32}\pi a^4.$

The reader should note in passing that the integral over A_1 gives the second moment of a circular area of radius a about a line through a point on the circumference perpendicular to its plane. The value is $\frac{3}{2}\pi a^4$.

Exercises 4:8

1. Show that the volume included between the elliptic paraboloid $2z = x^2/p + y^2/q$, the cylinder $x^2 + y^2 = a^2$, and the plane $z = 0$ is $\pi a^4(p + q)/8pq$.
2. A cylindrical hole of radius b $(<a)$ is bored symmetrically through a solid sphere of radius a. Find the volume of material removed from the sphere.
3. A solid of uniform density is bounded by the coordinate planes and the positive octant of the ellipsoid $x^2/a^2 + y^2/b^2 + z^2/c^2 = 1$. Find the moments of inertia of the solid about the coordinate axes.
4. Find the volume intercepted between the paraboloid $x^2 + y^2 = a(a - z)$ and the plane $z = 0$. Find also the position of the centroid of this volume.
5. Show in a diagram the region R of the x–y plane over which the integral

$$\int_0^a dx \int_0^{x^2/a} f(x, y)\, dy$$

is taken, assuming $a > 0$. A lamina covers the region R, and the density of the lamina at the point (x, y) is cx^2y, where c is a constant. Find the radius of gyration of the lamina about the x- and y-axes.

Miscellaneous Exercises 4

1 Sketch the area of integration of the integral

$$\int_0^a dx \int_{y_1}^{y_2} f(x, y)\, dy,$$

where $y_1 = a - x$, $y_2 = \sqrt{(a^2 - x^2)}$. Evaluate the integral when $f(x, y) = xy$

2. Evaluate $\iint x^2(b^2x^2 + a^2y^2)^{1/2}\,dx\,dy$ over the area enclosed by the ellipse $b^2x^2 + a^2y^2 = a^2b^2$.
3. Show in a diagram the region of integration of the integral

$$\int_0^{a\sqrt{2}} dx \int_x^{\sqrt{(a^2-x^2)}} xy(x+y)\,dy.$$

By changing to polar coordinates show that the value of the integral is $a^5/15$.
4. If A denotes the interior of the ellipse with equation

$$4x^2 + 9y^2 - 16x + 18y + 24 = 0,$$

evaluate

$$\iint_A (y^2 - 2y + 2)\,dx\,dy.$$

5. Sketch the area of integration of

$$\int_0^{4a} dy \int_{y^2/4a}^{y} \frac{y}{(4a-x)\sqrt{(4ax-y^2)}}\,dx,$$

where $a > 0$. Change the order of integration and evaluate the integral.
6. Show that by the transformation $x = u(1 - v)$, $y = uv$, the trapezium with vertices $(a, 0)$, $(b, 0)$, $(0, b)$, $(0, a)$ in the x–y plane is transformed into a rectangle in the u–v plane. Using this transformation, evaluate

$$\iint_A \frac{\sqrt{(x+y)}}{2\sqrt{[(x+y)^3+y]}}\,dx\,dy,$$

where A is the area of the trapezium.
Show also that

$$\int_0^b dx \int_0^{b-x} \frac{\sqrt{(x+y)}}{2\sqrt{[(x+y)^3+y]}}\,dy$$

has the value $\frac{1}{3}[(b^2 + 1)^{3/2} - b^3 - 1]$. [B.Sc.(Gen.)]

7. Show how to transform the double integral $\iint f(x, y)\,dx\,dy$ into generalised polar coordinates $x = ar\cos^k\theta$, $y = br\sin^k\theta$, where a, b, k are constants. Applying this transformation with a suitable value of k, find the volume bounded by the surface $(x^2/a^2 + y^2/b^2)^m + z/c = 1$, where $z > 0$ and m is a constant. [B.Sc.(Gen.)]

8. (i) Show that the volume common to the right circular cylinders $x^2 + y^2 = a^2$, $x^2 + z^2 = a^2$ is $16a^3/3$. [B.Sc.(Gen.)]
(ii) Show that the volume enclosed by the cylinders $x^2 + y^2 = 2ax$, $z^2 = 2ax$ is $128a^3/15$.

9. Find the volume of the cylinder $(x - a)^2 + y^2 = r^2$, $(a > r > 0)$, cut off by the hyperbolic paraboloid $xy = cz$, $(c > 0)$, in the octant where x, y, z are all positive. [B.Sc.(Spec.)]
10. Determine the volume in the positive octant enclosed between the surfaces $z = 0$ and $z = y^2 e^{-(x^2-y^2)}$.
11. Show that

$$\int_0^a dx \int_{(ax-x^2)^{1/2}}^{(a^2-x^2)^{1/2}} (a^2 - x^2 - y^2)^{1/2}\, dy$$

measures the volume of a part of the sphere $x^2 + y^2 + z^2 = a^2$. Illustrate your answer, showing clearly the section of the solid by the plane $z = 0$. Show that the value of the integral is $2a^3/9$.
12. Determine the volume which is bounded by the planes $z = 0$, $z = x + y$, $x = 0$, $y = a$, $y = 2a$ and the parabolic cylinder $y^2 = ax$.
13. Show that the volume enclosed between a sphere of radius a and centre the origin, and another sphere of radius $2a$ and centre at $(0, 0, 2a)$ is given by

$$\iint [(a^2 - r^2)^{1/2} - 2a + (4a^2 - r^2)^{1/2}] r\, dr\, d\theta,$$

where r, θ are plane polar coordinates in the x–y plane, and the limits of integration are appropriately chosen. Insert these limits and evaluate the integral. [B.Sc.(Gen.)]
14. Show that the volume enclosed by the surface $x^2 + y^2 = 4az$ and the plane $x + y + 2z = 4a$ is $25\pi a^3/2$.
15. Find the volume bounded by the surface $z = a^2 - x^2 - y^2$, the x–y plane and a circular cylinder which has the z-axis as one of its generators and diameter $2R(<a)$. [$z = a^2 - x^2 - y^2$ is the surface of revolution formed by rotating the parabola $z = a^2 - x^2$, $y = 0$ about the z-axis.] [B.Sc.(Gen.)]
16. Prove that the volume common to the sphere $x^2 + y^2 + z^2 = 1$ and the double cylinder $(x^2 + y^2)^2 = x^2 - y^2$ is $\frac{2}{9}[3\pi + 20 - 16\sqrt{2}]$.
17. A surface is given in spherical polar coordinates r, θ, ψ by the equation $r = a(1 + k \cos \theta)$, where $k \leqq 1$. Show that the volume enclosed by the surface is $\frac{4}{3}\pi a^3(1 + k^2)$.
18. Find the position of the centre of gravity of a uniform solid bounded by the coordinate planes and the positive octant of the ellipsoid $x^2/a^2 + y^2/b^2 + z^2/c^2 = 1$.
19. Show that the value of $\iiint(x^3 + y^3 + z^3)\, dx\, dy\, dz$ taken throughout the interior of the sphere $x^2 + y^2 + z^2 - 2a(x + y + z) + 2a^2 = 0$ is $32\pi a^6/5$.

20. Evaluate the integral

$$\iiint \frac{xy}{\sqrt{(z^2 - x^2 - y^2)}}\, dx\, dy\, dz$$

taken through the closed region in the positive octant bounded by the planes $x = 0$, $y = 0$, $z = 0$, the cone $z^2 = x^2 + y^2$, and the sphere $x^2 + y^2 + z^2 = a^2$. [B.Sc.(Gen.)]

21. A solid is bounded by the planes $x = 0, y = 0, z = 0, x/a + y/b + z/c = 1$. The density of the solid at the point (x, y, z) is $k(x/a + y/b + z/c + 1)^{-4}$ where k is constant. Show that the mass of the solid is $kabc/48$.

22. The axes of three circular cylinders each of radius a intersect one another at right angles. Show that the region which is interior to all three cylinders has a volume $8a^3(2 - \sqrt{2})$. If this region is bisected by a plane of symmetry perpendicular to one of the axes, find the distance from the plane of the centroid of each half. [B.Sc.(Eng.)]

23. Transform the integral $\iint[(x^2 - y^2)/(x^2 + y^2)]\, dx\, dy$ into an integral in polar coordinates r, θ, where $x = r\cos\theta$, $y = r\sin\theta$. Evaluate the integral over the region in the first quadrant bounded by an arc of the parabola $y^2 = 4(1 - x)$ and the coordinate axes. [B.Sc.(Eng.)]

24. Find the volume of the finite region bounded by the cone $z^2 = x^2 + y^2$ and the plane $x + y + 2z = 1$. (Use spherical polar coordinates.) [B.Sc.(Gen.)]

25. By applying the substitution $u = x$, $v = (x + y)/a$, show that

$$\iint_A e^{-[(x+y)/a]^2} \sinh\left(\frac{ax}{x + y}\right) dx\, dy = \tfrac{1}{2}a(\cosh a - 1),$$

where A is the infinite positive quadrant. [B.Sc.(Eng.)]

CHAPTER 5

SURFACE INTEGRALS

5:1 Introduction

We shall consider first the calculation of areas of curved surfaces and then integrals in which the field of integration is a curved surface. The *area* of a curved surface is one of the more difficult concepts of mathematics, and satisfactory definitions of it were not given until 1920. We shall follow here the treatment of De la Vallée Poussin, *Cours d'Analyse Infinitésimale*, Dover Reprint, with no attempt at justification.

A surface in three dimensions is conveniently given in terms of two parameters u, v, by three equations such as

$$x = f(u,v), \quad y = g(u,v), \quad z = h(u,v). \tag{1}$$

For example, $x = a \sin u \cos v$, $y = a \sin u \sin v$, $z = a \cos u$ are parametric equations of the sphere $x^2 + y^2 + z^2 = a^2$. Using suffixes to denote differentiation, we write

$$E = f_u^2 + g_u^2 + h_u^2, \quad G = f_v^2 + g_v^2 + h_v^2,$$
$$F = f_u f_v + g_u g_v + h_u h_v. \tag{2}$$

Then we *define* the element of curved surface area dS by

$$dS = \sqrt{(EG - F^2)}\, du\, dv, \tag{3}$$

and the curved surface area S by

$$S = \iint_A \sqrt{(EG - F^2)}\, du\, dv, \tag{4}$$

where A is the area in the u–v plane corresponding to the surface S, and the positive square root is taken. With certain restrictions on the functions f, g, h this provides a convenient treatment in general. (For further details, see E. G. Phillips, *A Course of Analysis*, Cambridge University Press, 1948.)

In what follows, we shall be concerned with surfaces which can be given by an equation of the form $z = \psi(x, y)$, where to any pair of values of x, y corresponds just one value of z. For example, $z = +\sqrt{(a^2 - x^2 - y^2)}$ is the equation of the hemisphere of radius a, centre the origin, for which $z > 0$. Similarly, $z = -\sqrt{(a^2 - x^2 - y^2)}$ gives the other hemisphere for which $z < 0$. Such an equation arises when it is possible to substitute in the equation $z = h(u, v)$ of (1) for u and v as functions of x and y. In our example of the sphere,

$$z^2 = a^2 \cos^2 u = a^2(1 - \sin^2 u),$$

or

$$z^2 = a^2 - x^2 - y^2,$$

giving the equation $z = \pm\sqrt{(a^2 - x^2 - y^2)}$. Here, equations (3) and (4) reduce to

$$\mathrm{d}S = \sqrt{(1 + z_x^2 + z_y^2)}\,\mathrm{d}x\,\mathrm{d}y, \tag{5}$$

$$S = \iint_A \sqrt{(1 + z_x^2 + z_y^2)}\,\mathrm{d}x\,\mathrm{d}y, \tag{6}$$

where A is the projection of S on the x–y plane. (For details, see Phillips, already quoted above.)

5:2 Calculation of surface areas

In applying 5:1(6) it is necessary to find A and $\sqrt{(1 + z_x^2 + z_y^2)}$ for the particular surface in question; then the problem reduces to a normal double integral. We illustrate with two examples.

EXAMPLE 1. *Find the surface area of the sphere* $x^2 + y^2 + z^2 = a^2$ *in the positive octant.*

The area A is found by putting $z = 0$ in the equation of the sphere, so A is the region inside the circle $x^2 + y^2 = a^2$ in the first quadrant, as shown in Fig. 23.

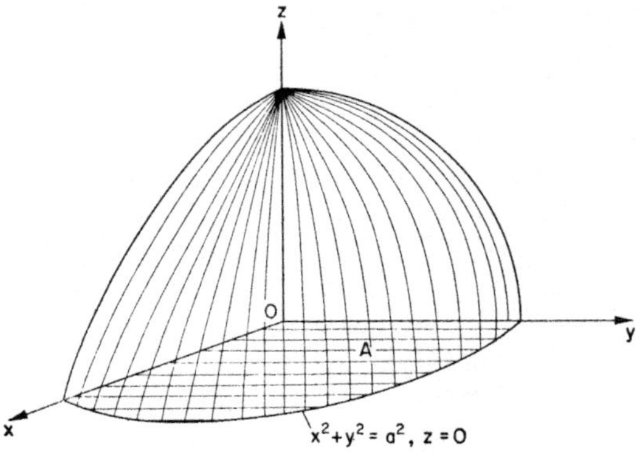

FIG. 23

Since $z^2 = a^2 - x^2 - y^2$, we have on differentiating partially,

$$2zz_x = -2x, \quad 2zz_y = -2y.$$

Thus $$z_x = -x/z, \qquad z_y = -y/z,$$

and $$\sqrt{(1 + z_x^2 + z_y^2)} = \sqrt{(1 + x^2/z^2 + y^2/z^2)}$$
$$= \sqrt{[(z^2 + x^2 + y^2)/z^2]}.$$

Therefore $$\sqrt{(1 + z_x^2 + z_y^2)} = a/z,$$

z being positive on the part of the sphere considered. So

$$S = \iint_A (a/z)\,\mathrm{d}x\,\mathrm{d}y$$
$$= a\iint_A [1/\sqrt{(a^2 - x^2 - y^2)}]\,\mathrm{d}x\,\mathrm{d}y.$$

To evaluate this integral, we change to polar coordinates ρ, ψ.

Then $$S = \iint_A [\rho/\sqrt{(a^2 - \rho^2)}]\,\mathrm{d}\rho\,\mathrm{d}\psi$$

$$= a\int_0^{\pi/2} d\psi \int_0^a [\rho/\sqrt{(a^2-\rho^2)}]\, d\rho$$

$$= a\int_0^{\pi/2} \Big[-\sqrt{(a^2-\rho^2)}\Big]_0^a d\psi.$$

So
$$S = \tfrac{1}{2}\pi a^2.$$

EXAMPLE 2. *Part of the surface $z^2 = 2xy$ is cut off by the planes $x = 0$, $x = a$, $y = 0$, $y = a$, where $a > 0$. Find its surface area.*

There are two parts of the surface $z^2 = 2xy$ to be considered, one for $z > 0$, the other for $z < 0$. The total area S is twice that for which $z > 0$. The area of integration A is the square bounded by the lines $x = 0$, $x = a$, $y = 0$, $y = a$ in the x–y plane.

Since $z^2 = 2xy$, we find on differentiation that

$$2zz_x = 2y, \quad 2zz_y = 2x.$$

So
$$z_x = y/z, \qquad z_y = x/z.$$

Thus
$$1 + z_x^2 + z_y^2 = 1 + y^2/z^2 + x^2/z^2$$
$$= (x^2 + y^2 + 2xy)/z^2$$
$$= (x + y)^2/2xy.$$

So
$$dS = [(x + y)/\sqrt{(2xy)}]\, dx\, dy,$$

where both x and y are positive. Hence

$$S = 2\iint_A [(x + y)/\sqrt{(2xy)}]\, dx\, dy$$

$$= \sqrt{2}\int_0^a dx \int_0^a [(x + y)/\sqrt{(xy)}]\, dy$$

$$= \sqrt{2}\int_0^a dx \int_0^a \left(\frac{\sqrt{x}}{\sqrt{y}} + \frac{\sqrt{y}}{\sqrt{x}}\right) dy$$

$$= \sqrt{2}\int_0^a \left[2\sqrt{(xy)} + \frac{2}{3}\frac{y\sqrt{y}}{\sqrt{x}}\right]_0^a dx$$

$$= 2\sqrt{2}\int_0^a \left[\sqrt{(ax)} + \frac{1}{3}a\sqrt{a}\,\frac{1}{\sqrt{x}}\right]\mathrm{d}x$$

$$= 2\sqrt{2}\left[\frac{2}{3}\sqrt{a}(x\sqrt{x}) + \frac{2}{3}a\sqrt{a}\sqrt{x}\right]_0^a.$$

Therefore $$S = 16\sqrt{2}a^2/3.$$

Exercises 5:2

1. An anchor ring is given by the parametric equations;

$$x = (2 + \sin\theta)\cos\phi, \quad y = (2 + \sin\theta)\sin\phi, \quad z = \cos\theta,$$

where $0 \leqq \theta < 2\pi$, $0 \leqq \phi < 2\pi$. Show that the surface area of the anchor ring is given by the double integral

$$\int_0^{2\pi}\mathrm{d}\phi\int_0^{2\pi}(2 + \sin\theta)\,\mathrm{d}\theta,$$

and evaluate this integral. Verify your result by finding the surface area of the ring by using Pappus's Theorem.

2. A sphere of radius a is intercepted by a circular cylinder of diameter a, the centre of the sphere lying on a generator of the cylinder. Show that the surface area of the sphere cut off by the cylinder is $2(\pi - 2)a^2$.

3. A closed surface is formed by rotating the cardioid $r = a(1 + \cos\theta)$ about the initial line $\theta = 0$. Taking the z-axis as the initial line and the origin as pole, show that the parametric equations of the surface are

$$x = a(1 + \cos\theta)\sin\theta\cos\phi, \quad y = a(1 + \cos\theta)\sin\theta\sin\phi,$$
$$z = a(1 + \cos\theta)\cos\theta.$$

Show that the total surface area is $32\pi a^2/5$. [B.Sc.(Eng.)]

4. Find the surface area of that part of the paraboloid $z = 9 - x^2 - y^2$ for which $z \geqq 0$.

5:3 Definition of $\iint F(x, y, z)\,\mathrm{d}x\,\mathrm{d}y$ over $z = f(x,y)$

There are two types of integral for which the field of integration is a curved surface. We define first $\iint F(x,y,z)\mathrm{d}x\mathrm{d}y$ over part of the surface $z = f(x,y)$, and then use this integral in § 5:4 in the definition of $\iint F(x,y,z)\,\mathrm{d}S$.

Suppose that S is part of the surface given by the equation $z = f(x,y)$; let the projection of S on the x–y plane be the area A

(see Fig. 24). At any point P on S draw the normal to the surface, in both directions away from the surface through P. Let γ be the angle the normal makes with the *positive* direction of Oz. Then

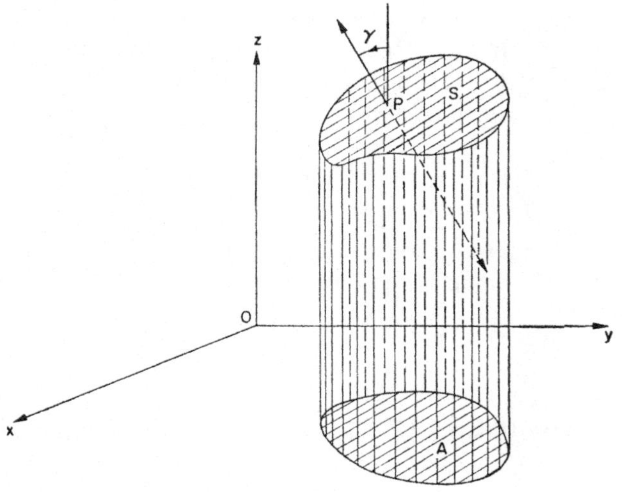

FIG. 24

for one direction of the normal γ is acute, for the other it is obtuse. That side of S for which γ is acute at all points is called the "upper" side of S. When γ is obtuse, we have the "lower" side of S.

We define $\iint F(x, y, z)\,\mathrm{d}x\,\mathrm{d}y$ over the "upper" side of S as

$$+\iint_A F(x, y, z)\,\mathrm{d}x\,\mathrm{d}y.$$

Similarly we define $\iint F(x, y, z)\,\mathrm{d}x\,\mathrm{d}y$ over the "lower" side of S by the integral

$$-\iint_A F(x, y, z)\,\mathrm{d}x\,\mathrm{d}y.$$

So integrals of this type are normal double integrals over an area in the x–y plane, but there is a sign to be determined in each case, and z must be replaced in the integrand by $f(x, y)$.

We can similarly define $\iint F(x,y,z)\,dy\,dz$ and $\iint F(x,y,z)\,dz\,dx$ over a surface S. For $\iint_S F(x,y,z)\,dy\,dz$ we must be able to write the equation of the surface in the form $x = f(y,z)$; then we need the angle α that the normal makes with the x-axis to determine sign, and the projection of S on the plane $x = 0$ to determine the area of integration. In the integrand we substitute $f(y,z)$ for x. Similar remarks apply to the definition of $\iint_S F(x,y,z)\,dz\,dx$.

EXAMPLE. *Evaluate the integral*

$$\iint[(1/x)\,dy\,dz + (1/y)\,dz\,dx + (1/z)\,dx\,dy]$$

over the outer surface of the ellipsoid $x^2/a^2 + y^2/b^2 + z^2/c^2 = 1$.

Let I be the value of the integral, and write

$$I = I_1 + I_2 + I_3,$$

where

$$I_1 = \iint(1/x)\,dy\,dz, \quad I_2 = \iint(1/y)\,dz\,dx, \quad I_3 = \iint(1/z)\,dx\,dy.$$

Consider now the evaluation of I_3. The ellipsoid is divided into halves by the plane $z = 0$. For $z > 0$, the outside of the ellipsoid is the "upper" side as defined earlier. For $z < 0$, the outside of the ellipsoid is the "lower" side. So we must attach a negative sign to the integral for $z < 0$. Thus

$$I_3 = \iint_{z>0}(1/z)\,dx\,dy - \iint_{z<0}(1/z)\,dx\,dy.$$

In the second integral on the right-hand side, $z < 0$ on the surface. Using this fact, and the symmetry of the surface over which we are integrating, we can write

$$I_3 = 2\iint_{z>0}(1/z)\,dx\,dy = 2\iint_A(1/z)\,dx\,dy,$$

the area A in this case being the area of the ellipse $x^2/a^2 + y^2/b^2 = 1$. On that part of the surface for which $z > 0$, we have at once, from the equation of the ellipsoid,

$$z = +c\sqrt{(1 - x^2/a^2 - y^2/b^2)}.$$

So the expression for I_3 becomes

$$I_3 = (2/c)\iint_A [1/\sqrt{(1 - x^2/a^2 - y^2/b^2)}]\,\mathrm{d}x\,\mathrm{d}y.$$

The evaluation of this integral is simplified if we first write

$$x = aX, \quad y = bY,$$

and then change X, Y to polar coordinates, the new area of integration A' being the unit circle $X^2 + Y^2 = 1$. We can, of course, combine these two transformations in one by writing

$$x = a\rho\cos\psi, \quad y = b\rho\sin\psi.$$

Then J, the Jacobian of the transformation, is $ab\rho$; so using 4:4(5),

$$\begin{aligned} I_3 &= \frac{2ab}{c}\iint_{A'} \frac{\rho\,\mathrm{d}\rho\,\mathrm{d}\psi}{\sqrt{(1-\rho^2)}} \\ &= \frac{2ab}{c}\int_0^{2\pi}\mathrm{d}\psi\int_0^1 \frac{\rho\,\mathrm{d}\rho}{\sqrt{(1-\rho^2)}} \\ &= \frac{2ab}{c}\,.\,2\pi\,.\left[-\sqrt{(1-\rho^2)}\right]_0^1. \end{aligned}$$

Therefore $$I_3 = 4\pi ab/c.$$

An exactly similar method yields $I_2 = 4\pi ca/b$, $I_1 = 4\pi bc/a$. Thus

$$I = I_1 + I_2 + I_3 = 4\pi(b^2c^2 + c^2a^2 + a^2b^2)/abc.$$

5:4 Definition of $\iint F(x, y, z)\,\mathrm{d}S$

We now define $\iint F(x, y, z)\,\mathrm{d}S$ taken over a part S of the surface $z = f(x, y)$, $\mathrm{d}S$ here being the element of (curved) surface area.

Let A be the projection of S on the x–y plane; then using the definition of dS in 5:1(5), we can write at once

$$\iint_S F(x, y, z)\,\mathrm{d}S = \iint_A F[x, y, f(x, y)]\sqrt{(1 + z_x^2 + z_y^2)}\,\mathrm{d}x\,\mathrm{d}y, \tag{1}$$

where, as usual, the positive square root sign is taken. It follows that $\iint_S F(x,y,z)\,\mathrm{d}S$ has the same value whether we integrate over the "upper" or "lower" side of S. The evaluation of $\iint F(x,y,z)\,\mathrm{d}S$ thus reduces to the evaluation of the double integral in x, y on the right of (1).

The reader should note that if the surface S is given instead by parametric equations of the form $x = f(u,v)$, $y = g(u,v)$, $z = h(u,v)$, then we make use of the definition of dS in 5:1(3) and

$$\iint_S F(x, y, z)\,\mathrm{d}S$$
$$= \iint_A F[f(u, v), g(u, v), h(u, v)]\sqrt{(E\,G - F^2)}\,\mathrm{d}u\,\mathrm{d}v, \tag{2}$$

where A is now the area in the u–v plane corresponding to S. So again we have an ordinary double integral to evaluate.

As an example, consider the evaluation of $\iint_S(x^2 + y^2 - 3z^2)\,\mathrm{d}S$, where S is the surface of the sphere $x^2 + y^2 + z^2 = 4$ above the plane $z = 0$. On the surface S, where $z^2 = 4 - x^2 - y^2$,

$$\mathrm{d}S = \sqrt{\left[1 + \left(\frac{x}{z}\right)^2 + \left(\frac{y}{z}\right)^2\right]}\mathrm{d}x\,\mathrm{d}y = \frac{1}{z}\sqrt{(x^2 + y^2 + z^2)}\,\mathrm{d}x\,\mathrm{d}y$$
$$= \frac{2}{z}\mathrm{d}x\,\mathrm{d}y = \frac{2}{\sqrt{(4 - x^2 - y^2)}}\mathrm{d}x\,\mathrm{d}y.$$

Therefore

$$I = \iint_S (x^2 + y^2 - 3z^2)\,\mathrm{d}S$$

$$I = \iint_A \left\{x^2 + y^2 - 3(4 - x^2 - y^2)\right\}\frac{2}{\sqrt{(4 - x^2 - y^2)}}\mathrm{d}x\,\mathrm{d}y,$$

where A is the circle $x^2 + y^2 = 4$. So

$$I = \iint_A 8(x^2 + y^2 - 3) \cdot \frac{1}{\sqrt{(4 - x^2 - y^2)}} \, dx \, dy.$$

Now change to polar coordinates $x = \rho \cos\psi$, $y = \rho \sin\psi$. Thus

$$I = \iint_A 8(\rho^2 - 3) \frac{\rho \, d\rho \, d\psi}{\sqrt{(4 - \rho^2)}}$$

$$= 8 \iint_A \left[-\rho\sqrt{(4 - \rho^2)} + \frac{\rho}{\sqrt{(4 - \rho^2)}} \right] d\rho \, d\psi$$

$$= 8 \int_0^{2\pi} d\psi \int_0^2 \left[-\rho\sqrt{(4 - \rho^2)} + \frac{\rho}{\sqrt{(4 - \rho^2)}} \right] d\rho$$

$$= 8 \cdot 2\pi \cdot \left[\tfrac{1}{3}(4 - \rho^2)^{3/2} - (4 - \rho^2)^{1/2}\right]_0^2$$

$$= -\tfrac{32}{3}\pi.$$

Exercises 5:4

1. Evaluate $\iint(zx \, dy \, dz + yz \, dz \, dx + z^2 \, dx \, dy)$ over that part of the surface of the sphere $x^2 + y^2 + z^2 = a^2$ which lies in the positive octant.
2. A finite solid is bounded by the paraboloid $x^2 + y^2 = 4z$ and the plane $z = 4$. Find the value of $\iint[y \, dz \, dx + (x + y + z) \, dx \, dy]$, taken over the surface of the solid.
3. Evaluate $\iint_S(x^4 + y^4 + z^4) \, dS$, where S is the surface of the sphere $x^2 + y^2 + z^2 = a^2$.
4. Evaluate $\iint_S(1/p) \, dS$, where S is the surface of the ellipsoid $x^2/a^2 + y^2/b^2 + z^2/c^2 = 1$, and p is the perpendicular distance from the origin to the tangent plane to the ellipsoid at the point where dS is situated.

5:5 Two theorems

Surface integrals occur in two very important theorems which we shall meet again later in vector form. We give them here in cartesian form, together with their proofs, which can be omitted by readers interested only in applications.

Gauss's Theorem (or Green's Theorem in space). *If $P(x,y,z)$, $Q(x,y,z)$, $R(x,y,z)$ are continuous functions defined throughout a* *volume V and on its bounding surface S, and if* $\partial P/\partial x$, $\partial Q/\partial y$, $\partial R/\partial z$ *are continuous in V and on S, then*

$$\iiint_V \left(\frac{\partial P}{\partial x} + \frac{\partial Q}{\partial y} + \frac{\partial R}{\partial z}\right) \mathrm{d}x\,\mathrm{d}y\,\mathrm{d}z$$

$$= \iint_S (P\,\mathrm{d}y\,\mathrm{d}z + Q\,\mathrm{d}z\,\mathrm{d}x + R\,\mathrm{d}x\,\mathrm{d}y), \qquad (1)$$

the surface integral being taken over the outer side of S.

Proof. We assume that any line parallel to any of the three axes meets S in not more than two points. [If this is not so, we subdivide S into regions for which it is true and add the separate integrals to get the integral over S.] We begin the proof by establishing the following result:

$$\iiint_V \frac{\partial R}{\partial z} \mathrm{d}x\,\mathrm{d}y\,\mathrm{d}z = \iint_S R(x, y, z)\,\mathrm{d}x\,\mathrm{d}y. \qquad (2)$$

Since S is a closed surface and we are integrating over the outside of it, it is clear that S can be treated as made up of two parts; a part S_1 on which we integrate over the "upper" side, as defined in § 5:3, and a part S_2 for which we integrate over the "lower" side. (The reader is referred back to the worked example in § 5:3 which involved integration over the outside of an ellipsoid; the parts S_1 and S_2 here were the halves of the surface for which $z > 0$ and $z < 0$ respectively.) We shall assume that S_1 has equation $z = f_1(x,y)$ and that S_2 has equation $z = f_2(x,y)$. In every case S_1 and S_2 are separated by a curve drawn on the surface S, and they both project into the same area A, say, on the x–y plane. Then

$$\iiint_V \frac{\partial R}{\partial z} \mathrm{d}x\,\mathrm{d}y\,\mathrm{d}z$$

$$= \iint_A \mathrm{d}x\,\mathrm{d}y \int_{z=f_2(x,y)}^{z=f_1(x,y)} \frac{\partial R}{\partial z} \mathrm{d}z$$

$$= \iint_A \Big[R(x, y, z) \Big]_{z=f_2(x,y)}^{z=f_1(x,y)} \mathrm{d}x\,\mathrm{d}y$$

$$= \iint_A R[x, y, f_1(x, y)]\,\mathrm{d}x\,\mathrm{d}y - \iint_A R[x, y, f_2(x, y)]\,\mathrm{d}x\,\mathrm{d}y. \tag{3}$$

In (3), by definition, $\iint_A R[x,y,f_1(x,y)]\,\mathrm{d}x\mathrm{d}y$ is a surface integral over the "upper" side of $z = f_1(x,y)$, and $-\iint R[x,y,f_2(x,y)]\,\mathrm{d}x\mathrm{d}y$ is a surface integral over the "lower" side of $z = f_2(x,y)$. Taken together these two integrals then give $\iint R(x,y,z)\mathrm{d}x\mathrm{d}y$ over the outside of S. So (2) is established. An exactly similar argument involving projections of parts of S on the planes $x = 0$ and $y = 0$ yields

$$\left.\begin{aligned} \iiint_V \frac{\partial P}{\partial x}\mathrm{d}x\,\mathrm{d}y\,\mathrm{d}z &= \iint_S P(x, y, z)\,\mathrm{d}y\,\mathrm{d}z, \\ \iiint_V \frac{\partial Q}{\partial y}\mathrm{d}x\,\mathrm{d}y\,\mathrm{d}z &= \iint_S Q(x, y, z)\,\mathrm{d}z\,\mathrm{d}x. \end{aligned}\right\} \tag{4}$$

Combining (2) and (4), we have Gauss's Theorem.

Gauss's Theorem is of great practical importance in that it can be used to replace a surface integral by a triple integral which is often much simpler to evaluate. As an example, suppose we use it to evaluate

$$\iint_S (x^3\,\mathrm{d}y\,\mathrm{d}z + y^3\,\mathrm{d}z\,\mathrm{d}x + z^3\,\mathrm{d}x\,\mathrm{d}y)$$

where S is the sphere of radius a and centre the origin. Here

$$P(x, y, z) = x^3, \quad Q(x, y, z) = y^3, \quad R(x, y, z) = z^3.$$

Gauss's Theorem gives

$$\iint_S (x^3\,\mathrm{d}y\,\mathrm{d}z + y^3\,\mathrm{d}z\,\mathrm{d}x + z^3\,\mathrm{d}x\,\mathrm{d}y)$$

$$= \iiint_V 3(x^2 + y^2 + z^2)\,\mathrm{d}x\,\mathrm{d}y\,\mathrm{d}z,$$

where V is the volume contained by the sphere S. In the triple integral we change to spherical polar coordinates. Then

$$\iiint_V 3(x^2+y^2+z^2)\,dx\,dy\,dz = 3\iiint_V r^4 \sin\theta\,dr\,d\theta\,d\psi$$

$$= 3\int_0^{2\pi} d\psi \int_0^{\pi} \sin\theta\,d\theta \int_0^a r^4\,dr$$

$$= \tfrac{12}{5}\pi a^5.$$

STOKES'S THEOREM. Suppose that a simple closed curve C in three-dimensional space is the bounding curve (or rim) of an open surface S. Suppose further that when C is traversed in the direction shown by the arrows (in Fig. 25) the outer side of the

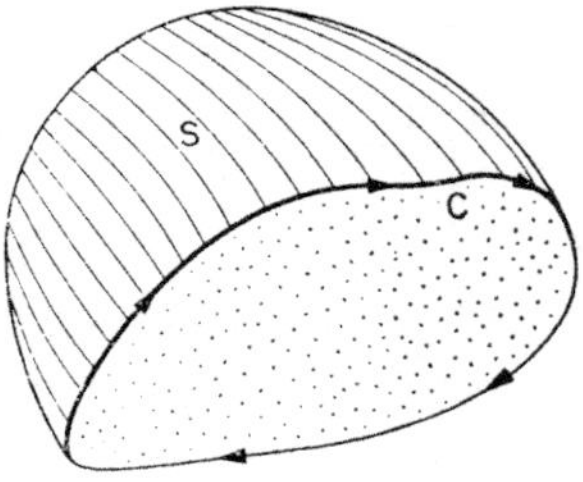

FIG. 25

surface S is to the left of C. Then Stokes's Theorem may be stated as follows:

If $P(x,y,z)$, $Q(x,y,z)$ $R(x,y,z)$ are continuous functions on both C and S, and if P_y, P_z, Q_x, Q_z, R_x, R_y are all continuous on C and S, then

$$\oint_C (P\,dx + Q\,dy + R\,dz)$$

$$= \iint_S [(R_y - Q_z)\,dy\,dz + (P_z - R_x)\,dz\,dx + (Q_x - P_y)\,dx\,dy], \quad (5)$$

the surface integral being taken over the outside of S and the line integral in the direction shown by the arrows.

Proof. It is clearly sufficient to establish the result

$$\oint_C P\,dx = \iint_S (P_z\,dz\,dx - P_y\,dx\,dy). \qquad (6)$$

Two further results will then follow by similar arguments, and we shall obtain Stokes's Theorem by addition. To prove (6), we assume that the surface S has an equation of the form $z = f(x,y)$.

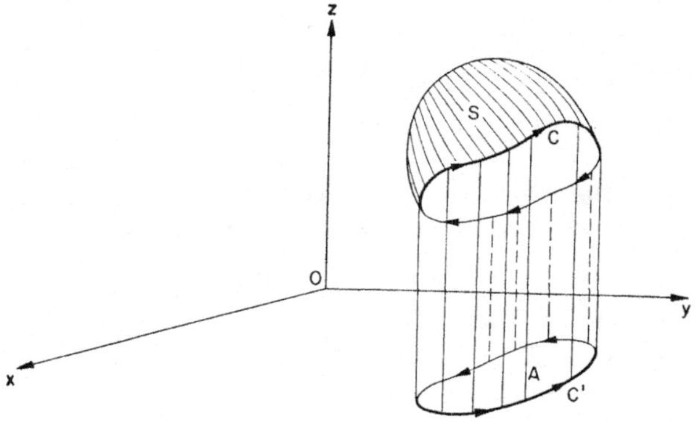

FIG. 26

In Fig. 26, let C' be the projection of C on the x–y plane, enclosing an area A which is the projection of S. If l, m, n are the direction cosines of the normal to the outer side of S, then

$$\frac{l}{z_x} = \frac{m}{z_y} = \frac{n}{-1}. \qquad (7)$$

Now dS is the element of area at a point $P(x,y,z)$ of S: it lies in the tangent plane to the surface at P. The projections of dS on the coordinate planes give us the elements of area in those planes. So

$$l\,dS = dy\,dz, \quad m\,dS = dz\,dx, \quad n\,dS = dx\,dy. \qquad (8)$$

Using (8), we can now write (6) as

$$\iint_S (P_z \,\mathrm{d}z\,\mathrm{d}x - P_y \,\mathrm{d}x\,\mathrm{d}y)$$

$$= \iint_S \left(\frac{\partial P}{\partial z} \cdot \frac{m}{n} - \frac{\partial P}{\partial y}\right) \mathrm{d}x\,\mathrm{d}y$$

$$= \iint_S \left(-\frac{\partial P}{\partial z} \cdot \frac{\partial z}{\partial y} - \frac{\partial P}{\partial y}\right) \mathrm{d}x\,\mathrm{d}y, \quad \text{from (7).} \tag{9}$$

On S, $z = f(x,y)$; so on S, $P(x,y,z) = P[x,y,f(x,y)]$. If we write $P_1(x,y)$ for $P[x,y,f(x,y)]$ on S, then

$$\frac{\partial P}{\partial z}\frac{\partial z}{\partial y} + \frac{\partial P}{\partial y} = \frac{\partial P_1}{\partial y}.$$

Now using the definition of a surface integral in § 5:3 and the result above, the surface integral in (9) may be rewritten as

$$\iint_S (P_z \,\mathrm{d}z\,\mathrm{d}x - P_y \,\mathrm{d}x\,\mathrm{d}y) = -\iint_A \frac{\partial P_1}{\partial y} \mathrm{d}x\,\mathrm{d}y. \tag{10}$$

In the line integral $\oint_C P\mathrm{d}x$, x and y take the same values on C as on C': hence

$$\oint_C P\,\mathrm{d}x = \oint_C P(x,y,z)\,\mathrm{d}x = \oint_{C'} P[x,y,f(x,y)]\,\mathrm{d}x$$

$$= \int_{C'} P_1(x,y)\,\mathrm{d}x.$$

We now apply Green's Theorem [4:6(1)] to this line integral:

$$\oint_C P\,\mathrm{d}x = \oint_{C'} P_1\,\mathrm{d}x = -\iint_A \frac{\partial P_1}{\partial y}\mathrm{d}x\,\mathrm{d}y. \tag{11}$$

Comparing (10) and (11) we find

$$\oint_C P\,dx = \iint_S (P_z\,dz\,dx - P_y\,dx\,dy).$$

A similar argument establishes separately

$$\oint_C Q\,dy = \iint_S (Q_x\,dx\,dy - Q_z\,dy\,dz),$$

$$\oint_C R\,dz = \iint_S (R_y\,dy\,dz - R_x\,dz\,dx).$$

Adding the three results, Stokes's Theorem is obtained.

We note that the necessary condition for a line integral to be independent of the path of integration is easily deduced from Stokes's Theorem. For if $\int_{AB} (P\,dx + Q\,dy + R\,dz)$ is independent of the path between two points A and B, then $\oint_C(P\,dx + Q\,dy + R\,dz) = 0$ for any closed curve C passing through A and B. This in turn implies that the surface integral to which $\oint_C(P\,dx + Q\,dy + R\,dz)$ is equal by Stokes's Theorem is zero for any open surface S of which C is the boundary; but there is an infinite number of such surfaces. The only possibility is thus that the surface integral be identically zero. Hence

$$R_y = Q_z,\ P_z = R_x,\ Q_x = P_y.$$

Alternative proofs of Gauss's and Stokes's Theorems will be given in §§ 6:10, 7:3, with examples of their use.

Exercises 5:5

1. Use Gauss's Theorem to evaluate

$$\iint_S (x^3\,dy\,dz + y^3\,dz\,dx + z^3\,dx\,dy)$$

where S is the surface of the cube bounded by the planes $x = \pm a$, $y = \pm a$, $z = \pm a$.

2. Show that

$$\iint_S (a^2x\,dy\,dz + b^2y\,dz\,dx + c^2z\,dx\,dy) = \tfrac{4}{3}\pi abc(a^2 + b^2 + c^2),$$

where S is the surface of the ellipsoid $x^2/a^2 + y^2/b^2 + z^2/c^2 = 1$.

3. Use Stokes's Theorem to evaluate the integral

$$\iint_S [(6y - 4z)\,dy\,dz + (6z - 4x)\,dz\,dx + (6x - 4y)\,dx\,dy],$$

where S is that part of the surface of the sphere $x^2 + y^2 - 2ax + az = 0$ for which $z \geqq 0$.

4. Evaluate directly

$$\iint_S [(x^2 + 2z)\,dy\,dz - 2xy\,dz\,dx + 3y^2\,dx\,dy],$$

where S is that part of the curved surface of the sphere $x^2 + y^2 + z^2 = a^2$ in the positive octant. Verify your result by using Stokes's Theorem.

Miscellaneous Exercises 5

1. Evaluate the integral $I = \iint_S z\,dS$, where S is that part of the surface $x = u\cos v$, $y = u\sin v$, $z = hv$ for which $0 \leqq u \leqq a$, $0 \leqq v \leqq \pi$. Find also the area A of this surface S, and show that $I/A = \frac{1}{2}h\pi$. [B.Sc.(Gen.)]

2. Find the area of that part of the surface of the sphere $x^2 + y^2 + z^2 = 1$ which lies inside the (double) cylinder $(x^2 + y^2)^2 = (x^2 - y^2)$. [B.Sc.(Gen.)]

3. Show that the equation in polar coordinates of a circle of radius a with centre on the initial line at a distance c from the origin is $r = c\cos\theta \pm \sqrt{(a^2 - c^2\sin^2\theta)}$.

   A right circular cylinder of radius a passes through a sphere of radius b $(b > a)$ in such a way that the surface area of the sphere cut off by the cylinder is in two separate parts. Find an expression for this surface area in the form of a double integral, distinguishing between the cases where the centre of the sphere lies inside and outside the cylinder.

   Evaluate the double integral (i) when the axis of the cylinder passes through the centre of the sphere, and (ii) when a generator of the cylinder passes through the centre of the sphere and $2a = b$. [B.Sc.(Gen.)]

4. A circular cylinder of unit radius penetrates a hemisphere of radius 2 units, and touches its circular base along a diameter. Prove that the volume and surface area of that part of the hemisphere lying outside the cylinder are $\frac{64}{9}$ and 16 units respectively.

5. A square hole of side $2\sqrt{2}$ inches is cut symmetrically through a sphere of radius 2 inches. Show that the area of the surface removed is $16\pi(\sqrt{2} - 1)$ square inches.

6. Evaluate

$$\iint_S (x^3y^3 + y^3z^3 + z^3x^3)\,dS,$$

where S is the surface intercepted on the upper half of the cone $z^2 = x^2 + y^2$ by the cylinder $x^2 + y^2 - 2y = 0$.

7. Evaluate

$$\iint_S [x^2 + y^2 + (a - z)^2]^{1/2}\,dS,$$

where S is that part of the surface of the sphere $x^2 + y^2 + z^2 = a^2$ lying in the positive octant.

8. A finite solid of volume V is bounded by the paraboloid $4z = x^2 + y^2$ and the plane $z = 4$. S is the surface of this solid. Verify by direct integration the equality of the integrals

$$\iint_S (x^2\,dy\,dz + y^2\,dz\,dx + z^2\,dx\,dy), \qquad \iiint_V 2(x + y + z)\,dx\,dy\,dz.$$

Deduce the value of the surface integral when the field of integration is the *curved* surface of the solid.

9. A solid is made up of a hemisphere of radius a together with a right circular cone of base radius a and height a, the plane bases of the hemisphere and the cone being coincident. If S denotes the total surface area of this solid, use Gauss's Theorem to evaluate

$$\iint_S (xz^2\,dy\,dz + z^3\,dx\,dy).$$

10. A solid in the positive octant is bounded by the planes $x = 0$, $y = 0$, $z = 0$ and the cylinders $x^2 + y^2 = a^2$, $x^2 + z^2 = a^2$. If S is the curved surface of this solid, evaluate

$$\iint_S [2(y + x)\,dz\,dx + (1 - 2z)\,dx\,dy].$$

Verify your result by applying Stokes's Theorem.

11. A solid in the positive octant is bounded by the planes $x = 0$, $y = 0$, $z = 0$, $z = 3$ and the cylinder $x^2 + y^2 = a^2$. If S is the curved surface of

this solid, use Stokes's Theorem to evaluate

$$\iint_S [(3x - 5xy)\,dy\,dz - 3y\,dz\,dx + (5yz - 1)\,dx\,dy].$$

Find also the value of this integral when the field of integration is the total surface area of this solid.

CHAPTER 6

VECTOR FIELD THEORY (1)

6:1 Introductory definitions

Vector Field Theory is an essential part of mathematics for the engineer and scientist. It provides a unified approach to many different branches of physics and mathematics; perhaps its greatest advantage is the concise and simple way in which it enables basic theory and formulae to be presented. We begin with some definitions of the types of function we shall be considering in this and the next chapter.

A *scalar point function* is one which has a definite value at each point of a given region in three-dimensional space. For example, the distance of the point (x, y, z) from the origin is such a function; $\sqrt{(x^2 + y^2 + z^2)}$ has a definite value at every point (x, y, z) of a given region. Or, given a volume of fluid, the temperature at each point of the fluid is a scalar point function. The term *scalar function of position* is also used for these functions. If $\phi(x, y, z)$ is the scalar point function corresponding to the point (x, y, z) of a region, we say that a *scalar field* ϕ has been defined for that region. It should be noticed that the value of ϕ is independent of the particular coordinates used: the distance of a point from the origin is *independent* of the directions of the axes; the temperature of a fluid clearly does not depend on the choice of the direction of the axes, nor on the particular origin chosen. Scalar point functions are always *invariant* under a rotation of the cartesian axes of reference; some are invariant under translation of the axes also, e.g. the temperature of a fluid.

A *vector point function* is a function which has direction as well as magnitude at any point. Thus to a point (x, y, z) corresponds a vector $\mathbf{V}(x, y, z)$. This is also known as a *vector function of position.* For example, the *position vector* $\mathbf{r}$ of the point (x, y, z) given by $\mathbf{r} = x\mathbf{i} + y\mathbf{j} + z\mathbf{k}$, where $\mathbf{i}, \mathbf{j}, \mathbf{k}$ are the unit vectors in the directions of the axes Ox, Oy, Oz respectively, is a vector point function. Or, the velocity at any point in a moving volume of fluid is a vector point function. A *vector field* $\mathbf{V}$ is defined in a region R when we know the vector point function $\mathbf{V}(x, y, z)$ for each point (x, y, z) of the region. The components of $\mathbf{V}(x, y, z)$ will change if we change the directions of the axes, but the magnitude and direction of $\mathbf{V}$ in space remain unaltered; for example, the velocity at a point in a fluid is unaffected by a rotation of the axes.

We have defined scalar and vector point functions as single-valued at each point of the field; this is usually the case in physical applications, though there are exceptions, such as the potential functions in hydrodynamics, but these will not concern us here. [See §§ 8:2 (viii) and 8:6 (ii) for examples of many-valued potential functions.]

In three dimensions the equation $\phi(x, y, z) = c$, where c is a constant, represents a surface. Hence the points of a region R, at which a scalar point function $\phi(x, y, z)$ takes the value c, lie on a certain surface. Similarly, the points where $\phi(x, y, z)$ takes the value c' lie on a second surface. The equation $\phi(x, y, z) = c$ defines a family of level surfaces; no two of them can intersect, since we have defined $\phi(x, y, z)$ as having a definite value at every point. Thus the spheres $x^2 + y^2 + z^2 = 1$, $x^2 + y^2 + z^2 = 2$ are two level surfaces of the family $x^2 + y^2 + z^2 = c$; clearly, no sphere of this system can intersect any other sphere of the system.

6:2 Gradient of a scalar field

Let $\phi(x, y, z) = c$ be a level surface. Take $P(x, y, z)$ and $Q(x + \mathrm{d}x, y + \mathrm{d}y, z + \mathrm{d}z)$ as neighbouring points on it (see Fig. 27). Since $\phi = c$ and c is a constant, the total differential $\mathrm{d}\phi$ is zero. So we have

$$\mathrm{d}\phi = 0 = \frac{\partial \phi}{\partial x}\mathrm{d}x + \frac{\partial \phi}{\partial y}\mathrm{d}y + \frac{\partial \phi}{\partial z}\mathrm{d}z. \qquad (1)$$

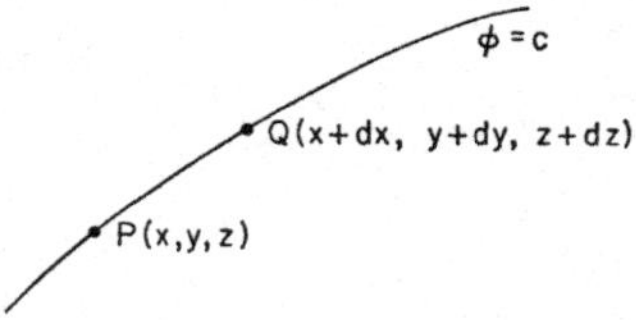

FIG. 27

It follows that the vectors

$$\frac{\partial\phi}{\partial x}\mathbf{i}+\frac{\partial\phi}{\partial y}\mathbf{j}+\frac{\partial\phi}{\partial z}\mathbf{k}, \tag{2}$$

$$\mathrm{d}x\,\mathbf{i}+\mathrm{d}y\,\mathbf{j}+\mathrm{d}z\,\mathbf{k}, \tag{3}$$

are at right angles; for their scalar product is zero by (1). But

$$\overrightarrow{PQ}=\mathrm{d}x\,\mathbf{i}+\mathrm{d}y\,\mathbf{j}+\mathrm{d}z\,\mathbf{k}.$$

Thus the vector

$$\frac{\partial\phi}{\partial x}\mathbf{i}+\frac{\partial\phi}{\partial y}\mathbf{j}+\frac{\partial\phi}{\partial z}\mathbf{k}$$

is at right angles to $\overrightarrow{PQ}$. If we now let Q approach P, then $\overrightarrow{PQ}$ becomes an arbitrary vector tangential to the surface at P. Hence the vector

$$\frac{\partial\phi}{\partial x}\mathbf{i}+\frac{\partial\phi}{\partial y}\mathbf{j}+\frac{\partial\phi}{\partial z}\mathbf{k}$$

is a vector which is normal to the level surface $\phi(x, y, z) = c$ at P.

Now consider the points $P(x, y, z)$, $R(x + \mathrm{d}x, y + \mathrm{d}y, z + \mathrm{d}z)$ where R is a point close to P on the normal to the surface at P (see Fig. 28).

Then

$$\frac{\mathrm{d}x}{\dfrac{\partial\phi}{\partial x}}=\frac{\mathrm{d}y}{\dfrac{\partial\phi}{\partial y}}=\frac{\mathrm{d}z}{\dfrac{\partial\phi}{\partial z}}.$$

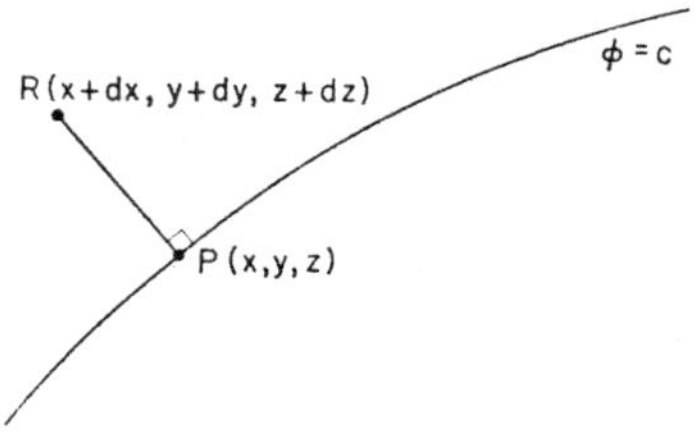

FIG. 28

Writing $\dfrac{\partial\phi}{\partial x} = \phi_x$, etc., for brevity, we have

$$\frac{dx}{\phi_x} = \frac{dy}{\phi_y} = \frac{dz}{\phi_z} = \frac{\sqrt{(dx^2 + dy^2 + dz^2)}}{\sqrt{(\phi_x^2 + \phi_y^2 + \phi_z^2)}}$$

$$= \frac{dn}{\sqrt{(\phi_x^2 + \phi_y^2 + \phi_z^2)}}, \qquad (4)$$

where dn is the length of the line-element PR.

Since ϕ is a function of x, y, and z, the increment $d\phi$ in ϕ due to the displacement $\overrightarrow{PR}$ is

$$d\phi = \frac{\partial\phi}{\partial x}dx + \frac{\partial\phi}{\partial y}dy + \frac{\partial\phi}{\partial z}dz.$$

So $$d\phi = (\phi_x^2 + \phi_y^2 + \phi_z^2)\frac{dn}{\sqrt{(\phi_x^2 + \phi_y^2 + \phi_z^2)}}, \text{ using (4)},$$

and $$d\phi = \sqrt{(\phi_x^2 + \phi_y^2 + \phi_z^2)}\,dn. \qquad (5)$$

Equation (5) gives the increment $d\phi$ in ϕ due to the displacement $PR = dn$, that is, due to a displacement along the normal. We denote the rate of change of ϕ in the direction of the normal by the partial derivative $\partial\phi/\partial n$; the partial derivative is used since, to specify the change in ϕ completely, we need the rates of change in three mutually perpendicular directions of which the normal is one. So from (5),

$$\frac{\partial\phi}{\partial n} = \sqrt{(\phi_x^2 + \phi_y^2 + \phi_z^2)}, \quad \mathrm{d}\phi = \frac{\partial\phi}{\partial n}\mathrm{d}n. \tag{6}$$

The magnitude of the vector

$$\frac{\partial\phi}{\partial x}\mathbf{i} + \frac{\partial\phi}{\partial y}\mathbf{j} + \frac{\partial\phi}{\partial z}\mathbf{k}$$

in (2) is, of course, $\sqrt{(\phi_x^2 + \phi_y^2 + \phi_z^2)}$. So using (6) we have

$$\frac{\partial\phi}{\partial x}\mathbf{i} + \frac{\partial\phi}{\partial y}\mathbf{j} + \frac{\partial\phi}{\partial z}\mathbf{k} = \frac{\partial\phi}{\partial n}\hat{\mathbf{n}},$$

where $\hat{\mathbf{n}}$ is the unit vector normal to the surface at P. The vector (2) is the *gradient of the scalar point function* ϕ. We denote it by grad ϕ. Thus

$$\text{grad}\,\phi = \frac{\partial\phi}{\partial x}\mathbf{i} + \frac{\partial\phi}{\partial y}\mathbf{j} + \frac{\partial\phi}{\partial z}\mathbf{k}. \tag{7}$$

As an example, suppose that our scalar point function ϕ is given by $\phi = x^3 + 2yx^2 + y^3 + z^2(x + y)$. Then

$$\text{grad}\,\phi = (3x^2 + 4xy + z^2)\mathbf{i} + (2x^2 + 3y^2 + z^2)\mathbf{j} + 2z(x + y)\mathbf{k},$$

and grad ϕ is itself a vector point function. Grad ϕ assumes a unique value at any point of the field. Thus at (1, 1, 1), grad $\phi = 8\mathbf{i} + 6\mathbf{j} + 4\mathbf{k}$; at (1, 0, 0), grad $\phi = 3\mathbf{i} + 2\mathbf{j}$, and so on.

Now consider points P, P' on the neighbouring level surfaces $\phi = c$, $\phi = c'$; then $(c - c')$ is small. Let PR be the normal at P to $\phi = c$, meeting $\phi = c'$ in R, as in Fig. 29. Write $PR = \mathrm{d}n$, $PP' = \mathrm{d}s$, $\angle P'PR = \theta$. Then

$$\mathrm{d}n = \mathrm{d}s\cos\theta. \tag{8}$$

The increment in ϕ due to the displacement $PR = \mathrm{d}n$ is $(\partial\phi/\partial n)\,\mathrm{d}n$. Similarly, the increment due to the displacement $PP' = \mathrm{d}s$ is

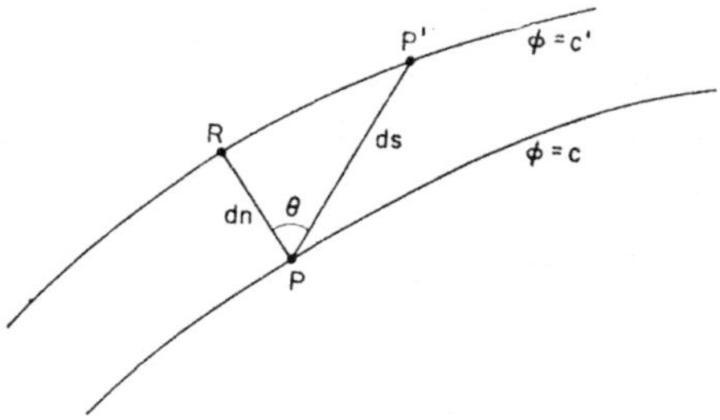

FIG. 29

$(\partial\phi/\partial s)\mathrm{d}s$. But in each case the increment must be $c' - c$, since we have moved from the surface $\phi = c$ to the surface $\phi = c'$. Therefore

$$\frac{\partial\phi}{\partial n}\mathrm{d}n = \frac{\partial\phi}{\partial s}\mathrm{d}s;$$

so

$$\frac{\partial\phi}{\partial n}\frac{\mathrm{d}n}{\mathrm{d}s} = \frac{\partial\phi}{\partial s}.$$

Using (8),

$$\frac{\partial\phi}{\partial n}\cos\theta = \frac{\partial\phi}{\partial s}. \tag{9}$$

$\partial\phi/\partial s$ is the *directional derivative* of ϕ in the direction given by $\mathrm{d}s$. It is greatest when $\cos\theta = 1$, that is in the direction of the normal. If $\hat{\mathbf{s}}$ is the unit vector in the direction given by $\mathrm{d}s$, that is in the direction of the vector $\overrightarrow{PP'}$, then $\cos\theta = \hat{\mathbf{s}}\,.\,\hat{\mathbf{n}}$. So (9) becomes

$$\frac{\partial\phi}{\partial s} = \frac{\partial\phi}{\partial n}(\hat{\mathbf{s}}\,.\,\hat{\mathbf{n}}) = \hat{\mathbf{s}}\,.\,\frac{\partial\phi}{\partial n}\hat{\mathbf{n}}.$$

Therefore

$$\frac{\partial\phi}{\partial s} = \hat{\mathbf{s}}\,.\,(\operatorname{grad}\phi). \tag{10}$$

Thus $\partial\phi/\partial s$ is the component of grad ϕ in the direction given by $\hat{\mathbf{s}}$.

Exercises 6:2

1. If $\phi = x^2 + y^2 + z^2$, find grad ϕ and |grad ϕ| at the point $(1, 1, \sqrt{2})$. Hence write down the unit vector normal to the sphere $x^2 + y^2 + z^2 = 4$ at the point $(1, 1, \sqrt{2})$.
2. If $\phi = 2yx^3 - z^2x$, find grad ϕ at (1, 1, 1) and show that the directional derivative of ϕ at (1, 1, 1) in the direction parallel to the vector $2\mathbf{i} - \mathbf{j} + \mathbf{k}$ is $\sqrt{6}$. In what direction at the point (1, 2, 3) is the directional derivative a maximum, and what is this maximum value?
3. The angle between two surfaces at a point common to both is defined to be the angle between the normals to the surfaces at that point. Write down vectors normal to the surfaces $3x^2 + 4y^2 - z^2 = 7$, $x^2 - y^2 = z$ at the point (2, 1, 3) and hence show that the acute angle between these surfaces at (2, 1, 3) is $\cos^{-1}(19/\sqrt{1281})$.

6:3 The vector operator ∇

We introduce now a most useful *vector operator*. Called del (or sometimes nabla) and written ∇, it is defined by

$$\nabla = \mathbf{i}\frac{\partial}{\partial x} + \mathbf{j}\frac{\partial}{\partial y} + \mathbf{k}\frac{\partial}{\partial z}. \tag{1}$$

Using it, we can write the gradient briefly as

$$\text{grad}\,\phi = \nabla\phi. \tag{2}$$

Thus $\nabla\phi$ denotes a vector whose components in the directions of **i, j, k** are $\partial\phi/\partial x$, $\partial\phi/\partial y$, $\partial\phi/\partial z$ respectively. It follows that 6:2(10) may now be written

$$\frac{\partial\phi}{\partial s} = \hat{\mathbf{s}}\,.\,\nabla\phi = \Delta\phi\,.\,\hat{\mathbf{s}}. \tag{3}$$

Writing $d\mathbf{s}$ for $ds\,\hat{\mathbf{s}}$, we obtain

$$d\phi = \frac{\partial\phi}{\partial s}ds = \nabla\phi\,.\,d\mathbf{s}. \tag{4}$$

Thus the increment in ϕ due to a vector increment $d\mathbf{s}$ is $\nabla\phi\,.\,d\mathbf{s}$; this is true for all vector increments $d\mathbf{s}$, whatever the direction of $\hat{\mathbf{s}}$.

It should be noted that $\nabla\phi$ forms one entity; ∇ and ϕ cannot be separated here.

Given the form of ϕ, the evaluation of $\nabla\phi$ is simply a matter of partial differentiation. For example, if $\phi = r^3$ where $r^2 = x^2 + y^2 + z^2$, then

$$\begin{aligned}\nabla\phi &= \left(\mathbf{i}\frac{\partial}{\partial x} + \mathbf{j}\frac{\partial}{\partial y} + \mathbf{k}\frac{\partial}{\partial z}\right)r^3 \\ &= 3r^2\left(\frac{\partial r}{\partial x}\mathbf{i} + \frac{\partial r}{\partial y}\mathbf{j} + \frac{\partial r}{\partial z}\mathbf{k}\right) \\ &= 3r^2\left(\frac{x}{r}\mathbf{i} + \frac{y}{r}\mathbf{j} + \frac{z}{r}\mathbf{k}\right) \\ &= 3r(x\mathbf{i} + y\mathbf{j} + z\mathbf{k}) \\ &= 3r\mathbf{r},\end{aligned}$$

where $\mathbf{r} = x\mathbf{i} + y\mathbf{j} + z\mathbf{k}$ is the position vector of the point (x, y, z). [A shorter method for this example is given later in 7:8(5).]

Suppose now that we have two scalar point functions ϕ_1 and ϕ_2. Then $\phi_1 + \phi_2$ is also a scalar point function and we can form its gradient:

$$\begin{aligned}&\nabla(\phi_1 + \phi_2) \\ &= \left(\mathbf{i}\frac{\partial}{\partial x} + \mathbf{j}\frac{\partial}{\partial y} + \mathbf{k}\frac{\partial}{\partial z}\right)(\phi_1 + \phi_2) \\ &= \mathbf{i}\left(\frac{\partial\phi_1}{\partial x} + \frac{\partial\phi_2}{\partial x}\right) + \mathbf{j}\left(\frac{\partial\phi_1}{\partial y} + \frac{\partial\phi_2}{\partial y}\right) + \mathbf{k}\left(\frac{\partial\phi_1}{\partial z} + \frac{\partial\phi_2}{\partial z}\right) \\ &= \left(\frac{\partial\phi_1}{\partial x}\mathbf{i} + \frac{\partial\phi_1}{\partial y}\mathbf{j} + \frac{\partial\phi_1}{\partial z}\mathbf{k}\right) + \left(\frac{\partial\phi_2}{\partial x}\mathbf{i} + \frac{\partial\phi_2}{\partial y}\mathbf{j} + \frac{\partial\phi_2}{\partial z}\mathbf{k}\right).\end{aligned}$$

Therefore $$\nabla(\phi_1 + \phi_2) = \nabla\phi_1 + \nabla\phi_2. \qquad (5)$$

Also the product $\phi_1\phi_2$ is a scalar point function. So

$$\begin{aligned}
\nabla(\phi_1\phi_2) &= \left(\mathbf{i}\frac{\partial}{\partial x}+\mathbf{j}\frac{\partial}{\partial y}+\mathbf{k}\frac{\partial}{\partial z}\right)(\phi_1\phi_2) \\
&= \mathbf{i}\frac{\partial}{\partial x}(\phi_1\phi_2)+\mathbf{j}\frac{\partial}{\partial y}(\phi_1\phi_2)+\mathbf{k}\frac{\partial}{\partial z}(\phi_1\phi_2) \\
&= \mathbf{i}\left(\phi_1\frac{\partial\phi_2}{\partial x}+\phi_2\frac{\partial\phi_1}{\partial x}\right)+\mathbf{j}\left(\phi_1\frac{\partial\phi_2}{\partial y}+\phi_2\frac{\partial\phi_1}{\partial y}\right)+ \\
&\qquad +\mathbf{k}\left(\phi_1\frac{\partial\phi_2}{\partial z}+\phi_2\frac{\partial\phi_1}{\partial z}\right) \\
&= \phi_1\left(\frac{\partial\phi_2}{\partial x}\mathbf{i}+\frac{\partial\phi_2}{\partial y}\mathbf{j}+\frac{\partial\phi_2}{\partial z}\mathbf{k}\right)+ \\
&\qquad +\phi_2\left(\frac{\partial\phi_1}{\partial x}\mathbf{i}+\frac{\partial\phi_1}{\partial y}\mathbf{j}+\frac{\partial\phi_1}{\partial z}\mathbf{k}\right).
\end{aligned}$$

Therefore
$$\nabla(\phi_1\phi_2) = \phi_1\nabla\phi_2+\phi_2\nabla\phi_1. \tag{6}$$

[Note that in (6) $\phi_1\nabla\phi_2$ and $\phi_2\nabla\phi_1$ are each vectors; $\phi_1\nabla\phi_2$ is the vector formed by multiplying the *vector* $\nabla\phi_2$ by the scalar ϕ_1.]

We see that the effect of operation by ∇ on a scalar function is exactly analogous to that of operation by $\mathrm{d}/\mathrm{d}x$ on a function of x. For if u, v are two functions of x, then the analogues of (5) and (6) are

$$\frac{\mathrm{d}}{\mathrm{d}x}(u+v) = \frac{\mathrm{d}u}{\mathrm{d}x}+\frac{\mathrm{d}v}{\mathrm{d}x}, \tag{7}$$

$$\frac{\mathrm{d}}{\mathrm{d}x}(uv) = u\frac{\mathrm{d}v}{\mathrm{d}x}+v\frac{\mathrm{d}u}{\mathrm{d}x}. \tag{8}$$

We can extend (5) and (6) to cover the sum and product of any number of scalar functions in the same way that (7) and (8) can be extended.

Exercises 6:3

1. Evaluate (i) ∇r^2, (ii) $\nabla(r^4 + r^7)$, (iii) $\nabla[f(r) + g(1/r)]$, where $r^2 = x^2 + y^2 + z^2$, and $f(r)$, $g(1/r)$ are arbitrary functions of r and $1/r$ respectively.
2. Evaluate (i) $\nabla(\mathbf{a} \cdot \mathbf{r})$, (ii) $\nabla[f(r)(\mathbf{a} \cdot \mathbf{r})]$, (iii) $\nabla[(\mathbf{a} \cdot \mathbf{r})(\mathbf{b} \cdot \mathbf{r})]$, where $\mathbf{r} = x\mathbf{i} + y\mathbf{j} + z\mathbf{k}$, $\mathbf{a} = a_1\mathbf{i} + a_2\mathbf{j} + a_3\mathbf{k}$, $\mathbf{b} = b_1\mathbf{i} + b_2\mathbf{j} + b_3\mathbf{k}$, a_1, a_2, a_3, b_1, b_2, b_3 being constants and $f(r)$ an arbitrary function of r.

6:4 Line integrals

Suppose that a particle is in motion along a given curve under the action of a force $\mathbf{F}$. Then the work done by the force in a displacement $\mathrm{d}\mathbf{r}$ from the position given by the position vector $\mathbf{r}$ is by definition $\mathbf{F} \cdot \mathrm{d}\mathbf{r}$. In Fig. 30, $\overrightarrow{OP} = \mathbf{r}$, $\overrightarrow{PQ} = \mathrm{d}\mathbf{r}$. If we write $\mathbf{F}$

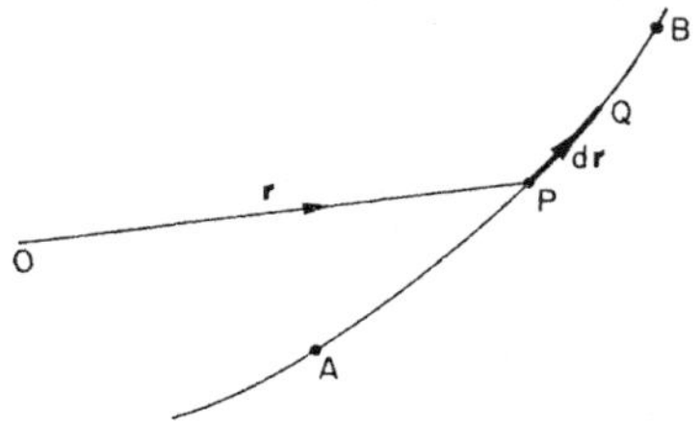

FIG. 30

and $\mathrm{d}\mathbf{r}$ in terms of their components as

$$\mathbf{F} = F_1\mathbf{i} + F_2\mathbf{j} + F_3\mathbf{k},$$

$$\mathrm{d}\mathbf{r} = \mathrm{d}x\mathbf{i} + \mathrm{d}y\mathbf{j} + \mathrm{d}z\mathbf{k},$$

then the work done in the displacement $\mathrm{d}\mathbf{r}$ is $(F_1\mathrm{d}x + F_2\mathrm{d}y + F_3\mathrm{d}z)$, where in general F_1, F_2, F_3 are functions of x, y, z, and $\mathrm{d}x$, $\mathrm{d}y$, $\mathrm{d}z$ are not all arbitrary, since they refer to displacements on a curve.

The total work done in a displacement from A to B is found by integration; it is $\int_{AB}(F_1\mathrm{d}x + F_2\mathrm{d}y + F_3\mathrm{d}z)$, which is a line integral of the type met with in § 3:4. So in this example,

$$\int_{AB} (F_1\,dx + F_2\,dy + F_3\,dz) = \int_{AB} \mathbf{F}\,.\,d\mathbf{r}.$$

In general, a curve in three dimensions is given parametrically by $x = x(t)$, $y = y(t)$, $z = z(t)$. The position of any point on the curve can be defined by the vector $\mathbf{r}(t)$, where

$$\mathbf{r}(t) = x(t)\mathbf{i} + y(t)\mathbf{j} + z(t)\mathbf{k}.$$

If now $\mathbf{V}(x,y,z)$ is a vector point function with components $V_1(x,y,z)$, $V_2(x,y,z)$, $V_3(x,y,z)$, then

$$\int_{AB} \mathbf{V}\,.\,d\mathbf{r} = \int_{AB} (V_1\,dx + V_2\,dy + V_3\,dz), \qquad (1)$$

or, in terms of t,

$$\int_{AB} \mathbf{V}\,.\,d\mathbf{r} = \int_{\alpha}^{\beta} \left(V_1\frac{dx}{dt} + V_2\frac{dy}{dt} + V_3\frac{dz}{dt}\right)dt, \qquad (2)$$

where $t = \alpha$ at A and $t = \beta$ at B.

From the properties of line integrals two results follow at once:

(i) $$\int_{AB} \mathbf{V}\,.\,d\mathbf{r} = \int_{AP} \mathbf{V}\,.\,d\mathbf{r} + \int_{PB} \mathbf{V}\,.\,d\mathbf{r},$$

where P is any point of the curve AB,

(ii) $$\int_{AB} \mathbf{V}\,.\,d\mathbf{r} = -\int_{BA} \mathbf{V}\,.\,d\mathbf{r}.$$

If C is a closed curve, we denote the line integral of $\mathbf{V}$ round C in an anti-clockwise direction by

$$\oint_C \mathbf{V}\,.\,d\mathbf{r}.$$

In two dimensions,

$$\int_{AB} \mathbf{V}\,.\,d\mathbf{r} = \int_{AB} (V_1\,dx + V_2\,dy).$$

EXAMPLE 1. *Evaluate* $\int_{AB} \mathbf{V} . \mathbf{dr}$, *if* $\mathbf{V} = 3xy\mathbf{i} - 5z\mathbf{j} + 10x\mathbf{k}$, *and* A, B *are points* (2, 2, 1), (5, 8, 8) *respectively on the curve* $x = 1 + t^2$, $y = 2t^2$, $z = t^3$.

From the values of **V** and **dr** we have

$$\mathbf{V} . \mathbf{dr} = 3xy\,\mathrm{d}x - 5z\,\mathrm{d}y + 10x\,\mathrm{d}z.$$

Now express the right-hand side in terms of t.

$$\int_{AB} \mathbf{V} . \mathbf{dr}$$

$$= \int_1^2 [3(1 + t^2) . 2t^2 . 2t . \mathrm{d}t - 5t^3 . 4t . \mathrm{d}t + 10(1 + t^2) . 3t^2 . \mathrm{d}t]$$

$$= \int_1^2 (12t^5 + 10t^4 + 12t^3 + 30t^2)\,\mathrm{d}t.$$

Therefore $$\int_{AB} \mathbf{V} . \mathbf{dr} = 303, \text{ on integration.}$$

EXAMPLE 2. *Evaluate* $\oint_C \mathbf{V} . \mathbf{dr}$, *where* $\mathbf{V} = (2y^2 - x^2)\mathbf{i} + (3x^2 - y^2)\mathbf{j}$, *and* C *is the circle* $(x - a)^2 + y^2 = a^2$.

Here,

$$\oint_C \mathbf{V} . \mathbf{dr} = \oint_C [(2y^2 - x^2)\,\mathrm{d}x + (3x^2 - y^2)\,\mathrm{d}y].$$

The value of this line integral is $6\pi a^3$; see § 4:6.

The integral $\oint_C \mathbf{V} . \mathbf{dr}$ is known as the circulation in the circuit C (this term arises in fluid dynamics, where V is a velocity; see § 8:2 (vii)).

Exercises 6:4

1. Evaluate $\int_{AB} \mathbf{V} . \mathbf{dr}$, if $\mathbf{V} = (3x^2y + zy^2)\mathbf{i} + (x^3 + 2xyz)\mathbf{j} + y^2x\mathbf{k}$, and A, B are the points (1, 2, 3), (2, 4, 12) respectively on the curve $x = t$, $y = 2t$, $z = 3t^2$.
2. If $\mathbf{V} = 2x^2\mathbf{i} + 3y\mathbf{j} + z^2\mathbf{k}$, $\mathbf{r} = x\mathbf{i} + y\mathbf{j} + z\mathbf{k}$, show that $\mathbf{V} \times \mathbf{dr} = (3y\,\mathrm{d}z - z^2\,\mathrm{d}y)\mathbf{i} + (z^2\,\mathrm{d}x - 2x^2\,\mathrm{d}z)\mathbf{j} + (2x^2\,\mathrm{d}y - 3y\,\mathrm{d}x)\mathbf{k}$. Hence find the value of $\oint_C \mathbf{V} \times \mathbf{dr}$, when C is the circle $z = 0$, $x^2 + y^2 = 4$.

6:5 Irrotational vector fields

In a general case, the circulation may be zero for a number of particular circuits in a vector field, but will otherwise be non-zero. If, however, the circulation is zero for every circuit in a vector field, then this implies a restriction on the vector field $\mathbf{V}$; for it implies that the line integral

$$\int_{AB} (V_1 \mathrm{d}x + V_2 \mathrm{d}y + V_3 \mathrm{d}z)$$

is independent of the path of integration, as we shall show shortly.

A vector field is said to be *irrotational*, or *conservative*, when the circulation is zero for every circuit in the field.

The reader will realise that analytically the condition for $\mathbf{V}$ to be irrotational is

$$\frac{\partial V_1}{\partial y} = \frac{\partial V_2}{\partial x}, \quad \frac{\partial V_2}{\partial z} = \frac{\partial V_3}{\partial y}, \quad \frac{\partial V_3}{\partial x} = \frac{\partial V_1}{\partial z} \qquad \text{[cf. 3:4(3)].}$$

We shall return to this condition in § 7:3. Meanwhile, we prove two important results:

(i) an irrotational vector field $\mathbf{V}$ can always be expressed as the gradient of a scalar point function ϕ,

(ii) if ϕ is a scalar point function, then grad ϕ is irrotational.

ϕ is known as the *scalar potential.*

To prove (i), we begin with an irrotational vector field $\mathbf{V}$; that is we know that $\oint_C \mathbf{V} \cdot \mathrm{d}\mathbf{r} = 0$ for all C in the field. Suppose we choose two points A, B in the field. Let P, Q be points on a curve C through A and B, as shown in Fig. 31. Then

$$\oint_C \mathbf{V} \cdot \mathrm{d}\mathbf{r} = \int_{APB} \mathbf{V} \cdot \mathrm{d}\mathbf{r} + \int_{BQA} \mathbf{V} \cdot \mathrm{d}\mathbf{r} = 0.$$

So

$$\int_{APB} \mathbf{V} \cdot \mathrm{d}\mathbf{r} = \int_{AQB} \mathbf{V} \cdot \mathrm{d}\mathbf{r}. \qquad (1)$$

But C is any curve through A and B. So (1) shows that $\int_{AB} \mathbf{V} \cdot \mathrm{d}\mathbf{r}$

is independent of the path of integration. Now let us treat A as fixed and allow B to vary. Then the value of $\int_{AB} \mathbf{V} \,.\, \mathrm{d}\mathbf{r}$ depends

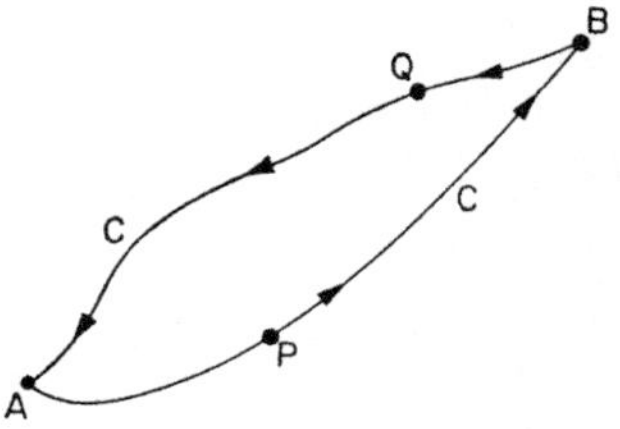

FIG. 31

on the position of B only. But $\int_{AB} \mathbf{V} \,.\, \mathrm{d}\mathbf{r}$ is a scalar, and hence it must be a scalar point function depending on B, $\phi(B)$ say. Thus we write

$$\phi(B) = \int_{AB} \mathbf{V} \,.\, \mathrm{d}\mathbf{r}.$$

If B' is another point in the field near B,

$$\phi(B') = \int_{AB'} \mathbf{V} \,.\, \mathrm{d}\mathbf{r}.$$

So
$$\phi(B') - \phi(B) = \int_{AB'} \mathbf{V} \,.\, \mathrm{d}\mathbf{r} - \int_{AB} \mathbf{V} \,.\, \mathrm{d}\mathbf{r}$$
$$= \int_{BB'} \mathbf{V} \,.\, \mathrm{d}\mathbf{r}.$$

Now let BB' tend to zero. Then $\phi(B') - \phi(B) \to \mathrm{d}\phi$, and we have

$$\mathrm{d}\phi = \mathbf{V} \,.\, \mathrm{d}\mathbf{r}. \tag{2}$$

But we know [6:3(4)] that if ϕ is a scalar point function then the increment in ϕ due to a vector increment $\mathrm{d}\mathbf{r}$ is given by

$$\mathrm{d}\phi = \nabla\phi \,.\, \mathrm{d}\mathbf{r}. \tag{3}$$

Combining (2) and (3),

$$(\mathbf{V} - \nabla\phi) \,.\, \mathrm{d}\mathbf{r} = 0. \tag{4}$$

In (4) **dr** is arbitrary, since **dr** is a vector displacement on any curve through B. So the only possible solution to (4) is that $(\mathbf{V} - \nabla\phi)$ be zero; hence

$$\mathbf{V} = \nabla\phi.$$

This establishes the result; a scalar potential function ϕ exists.

To prove (ii), we begin with a given scalar ϕ. Write $\nabla\phi = \mathbf{V}$, and let C be any circuit in the vector field $\mathbf{V}$. Then

$$\oint_C \mathbf{V} \,.\, d\mathbf{r} = \oint_C \nabla\phi \,.\, d\mathbf{r}$$

$$= \oint_C d\phi, \quad \text{by 6:3(4).}$$

Now $\oint_C d\phi$ gives the change in ϕ when C is traversed in an anti-clockwise direction, and this change is zero, since ϕ is a scalar point function. Hence $\oint_C \mathbf{V} \,.\, d\mathbf{r} = 0$, and so $\nabla\phi$ is irrotational.

As an example, consider the vector field given by

$$\mathbf{V} = yz\mathbf{i} + zx\mathbf{j} + xy\mathbf{k}.$$

This will be irrotational if we can find a scalar function such that $\nabla\phi = \mathbf{V}$. Now

$$\nabla\phi = \frac{\partial\phi}{\partial x}\mathbf{i} + \frac{\partial\phi}{\partial y}\mathbf{j} + \frac{\partial\phi}{\partial z}\mathbf{k}.$$

So we require $\partial\phi/\partial x = yz$, $\partial\phi/\partial y = zx$, $\partial\phi/\partial z = xy$. These equations are clearly satisfied if we take

$$\phi = xyz + \text{constant}.$$

Hence the given field is irrotational. The arbitrary constant in the expression for ϕ creates no difficulty here; for any particular chosen value of the constant the change in ϕ round any circuit in the field is zero, and in a given problem it will be possible to fix the constant according to known conditions.

Exercises 6:5

1. Show that $\mathbf{V} = (3x^2y + zy^2)\mathbf{i} + (x^3 + 2xyz)\mathbf{j} + y^2x\mathbf{k}$ is irrotational and find the scalar potential function ϕ. Hence evaluate the line integral in Exercises 6:4, 1.
2. Which of the following vector fields are irrotational? (i) $r^2\mathbf{r}$, (ii) $(\mathbf{a} \cdot \mathbf{r})\mathbf{r}$, (iii) $(r^2 + 2r)\mathbf{a}$, (iv) $f(r)\mathbf{r}$, where $\mathbf{r} = x\mathbf{i} + y\mathbf{j} + z\mathbf{k}$, $\mathbf{a}$ is a constant vector, and $f(r)$ is an arbitrary function of r. Find also the scalar potential function ϕ in (i).

6:6 Surface integrals

Suppose that we have a surface S in three dimensions. At any point P of this surface we can form the element of surface area $\mathrm{d}S$. If $\hat{\mathbf{n}}$ is the unit outward vector normal to the surface at P (i.e. the normal to the "upper" side of S as defined in § 5:3), we define the vector element of area at P by

$$\mathrm{d}\mathbf{S} = \mathrm{d}S\hat{\mathbf{n}}. \tag{1}$$

Thus $\mathrm{d}\mathbf{S}$ is a vector in the direction of the normal to the surface at P. We can now interpret the surface integral $\int_S \mathbf{V} \cdot \mathrm{d}\mathbf{S}$; for, using (1),

$$\int_S \mathbf{V} \cdot \mathrm{d}\mathbf{S} = \int_S (\mathbf{V} \cdot \hat{\mathbf{n}})\,\mathrm{d}S. \tag{2}$$

In (2), $\mathbf{V} \cdot \hat{\mathbf{n}}$ is a scalar, and so $\int_S (\mathbf{V} \cdot \hat{\mathbf{n}})\mathrm{d}S$ is a surface integral such as we have already discussed in Chapter 5. If the surface S is one whose equation can be put in the form $z = f(x,y)$, then $\int_S (\mathbf{V} \cdot \hat{\mathbf{n}})\,\mathrm{d}S$ can be interpreted as in § 5:4, by replacing $\mathrm{d}S$ by $\sqrt{(1 + z_x^2 + z_y^2)}\,\mathrm{d}x\,\mathrm{d}y$. Alternatively, using 5:5(8), we may write

$$\mathrm{d}\mathbf{S} = \mathrm{d}y\,\mathrm{d}z\mathbf{i} + \mathrm{d}z\,\mathrm{d}x\mathbf{j} + \mathrm{d}x\,\mathrm{d}y\mathbf{k}.$$

Then

$$\int_S \mathbf{V} \cdot \mathrm{d}\mathbf{S} = \int_S (V_1\,\mathrm{d}y\,\mathrm{d}z + V_2\,\mathrm{d}z\,\mathrm{d}x + V_3\,\mathrm{d}x\,\mathrm{d}y), \tag{3}$$

where $\mathbf{V} = V_1\mathbf{i} + V_2\mathbf{j} + V_3\mathbf{k}$. The surface integral (3) is now of the type considered in § 5:3.

We next introduce the *flux* of a vector. At any point P, say, in a vector field $\mathbf{V}$ draw a unit vector $\hat{\mathbf{n}}$ in any chosen direction. Construct the plane through P perpendicular to $\hat{\mathbf{n}}$, and in this plane take the element of area dS surrounding P (see Fig. 32). Then

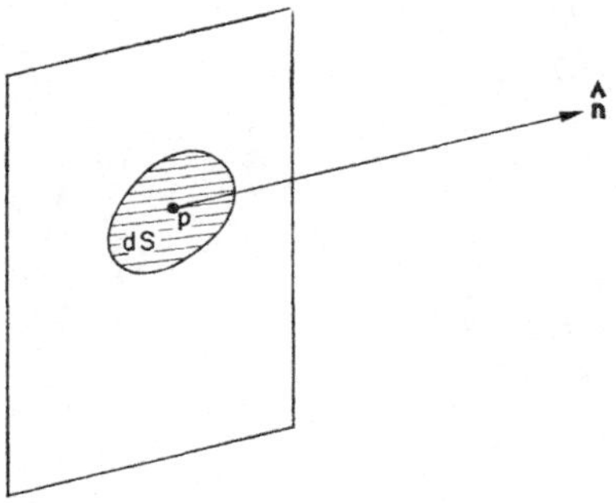

FIG. 32

$(\mathbf{V} \cdot \hat{\mathbf{n}})\, dS$ is the *flux* of V through the element of area dS in the direction of $\hat{\mathbf{n}}$. This can be written $\mathbf{V} \cdot d\mathbf{S}$, where $d\mathbf{S}$ is the vector element of area at P. The extension to a finite surface S follows by integration; the *flux* of $\mathbf{V}$ through a finite surface area S is defined to be $\int_S \mathbf{V} \cdot d\mathbf{S}$.

We see that the flux of a vector is itself a scalar quantity, since $\mathbf{V} \cdot d\mathbf{S}$ is a scalar. The concept of flux occurs naturally in subjects such as fluid dynamics, where $\mathbf{V}$ is the velocity of a fluid, and in electrostatics where $\mathbf{V}$ is the intensity of the field; see § 8:2 (iii).

6:7 The divergence of a vector field

Suppose we have a point P in a vector field $\mathbf{V}$ surrounded by a small closed surface S; denote the volume enclosed by S by δv. Then the flux of $\mathbf{V}$ through the surface S is δF, where

$$\delta F = \int_S \mathbf{V} \cdot d\mathbf{S}. \qquad (1)$$

We have noted already that flux is a scalar. We now introduce a new scalar point function, the *divergence* of the vector point function **V**. We denote it briefly by div **V**, and define it at the point P by the equation

$$\operatorname{div} \mathbf{V} = \lim_{\delta v \to 0} \frac{\delta F}{\delta v}. \tag{2}$$

From (2) it follows that

$$\operatorname{div} \mathbf{V} = \lim_{\delta v \to 0} \left[\frac{1}{\delta v} \int_S (\mathbf{V} \cdot d\mathbf{S}) \right]. \tag{3}$$

The integral form in (3) is very useful in theoretical work; for practical applications we need the analytical form, which we shall now establish. This is

$$\operatorname{div} \mathbf{V} = \frac{\partial V_1}{\partial x} + \frac{\partial V_2}{\partial y} + \frac{\partial V_3}{\partial z}. \tag{4}$$

(Those readers interested only in applications may prefer the alternative definition of div **V** noted in § 6:8.)

To establish (4) from our integral definition we have to evaluate the limit in (3). We take the surface S as a rectangular parallelepiped of sides dx, dy, dz with the point $P(x, y, z)$ at its centre, as shown in Fig. 33. We have to find $\lim_{\delta v \to 0} \delta F/\delta v$, and here

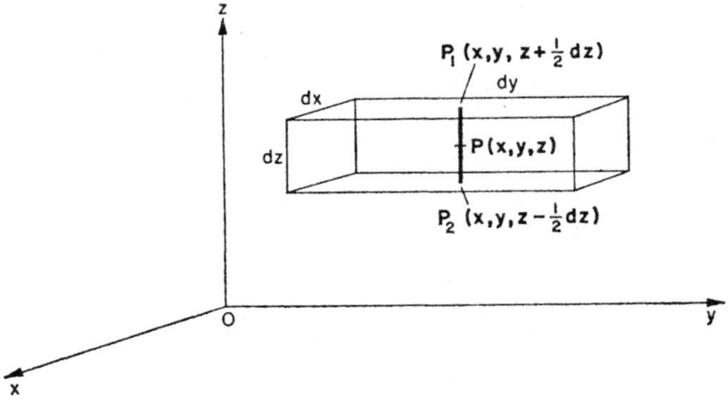

FIG. 33

$\delta v = \mathrm{d}x\,\mathrm{d}y\,\mathrm{d}z$. Also $\delta F = \int_S \mathbf{V}\,.\,\mathrm{d}\mathbf{S}$, and the integral over S breaks up into six parts, one corresponding to each face of S. Since each face is an element of area, the flux through each face is simply given by $\mathbf{V}\,.\,\mathrm{d}\mathbf{S}$ for each face. Consider the two faces parallel to $z = 0$. In each case $\mathrm{d}S = \mathrm{d}x\,\mathrm{d}y$; on the upper face $\mathrm{d}\mathbf{S} = \mathrm{d}x\,\mathrm{d}y\mathbf{k}$, and on the lower face $\mathrm{d}\mathbf{S} = -\,\mathrm{d}x\,\mathrm{d}y\mathbf{k}$, since the outward normals to these faces are $\mathbf{k}$, $-\mathbf{k}$ respectively. If we denote by P_1 and P_2 the points $(x, y, z + \frac{1}{2}\mathrm{d}z)$, $(x, y, z - \frac{1}{2}\mathrm{d}z)$ respectively, then the flux through the upper face is, by definition,

$$\begin{aligned}\mathbf{V}(P_1)\,.\,\mathrm{d}\mathbf{S} &= \mathbf{V}(x, y, z + \tfrac{1}{2}\mathrm{d}z)\,.\,\mathrm{d}x\,\mathrm{d}y\mathbf{k}\\ &= V_3(x, y, z + \tfrac{1}{2}\mathrm{d}z)\,\mathrm{d}x\,\mathrm{d}y. \qquad (5)\end{aligned}$$

For the lower face, the flux is

$$\begin{aligned}\mathbf{V}(P_2)\,.\,\mathrm{d}\mathbf{S} &= \mathbf{V}(x, y, z - \tfrac{1}{2}\mathrm{d}z)\,.\,-\,\mathrm{d}x\,\mathrm{d}y\mathbf{k}\\ &= -\,V_3(x, y, z - \tfrac{1}{2}\mathrm{d}z)\,\mathrm{d}x\,\mathrm{d}y. \qquad (6)\end{aligned}$$

Combining (5) and (6), the flux δF_z for the two faces is

$$\delta F_z = [V_3(x, y, z + \tfrac{1}{2}\mathrm{d}z) - V_3(x, y, z - \tfrac{1}{2}\mathrm{d}z)]\,\mathrm{d}x\,\mathrm{d}y.$$

If we now expand by Taylor's Theorem, we obtain

$$\begin{aligned}\delta F_z = &\left[V_3(x, y, z) + \tfrac{1}{2}\mathrm{d}z\,.\,\frac{\partial}{\partial z}V_3(x, y, z) + \right.\\ &\qquad\left. + \text{ higher order terms}\right]\mathrm{d}x\,\mathrm{d}y\, -\\ &-\left[V_3(x, y, z) - \tfrac{1}{2}\mathrm{d}z\,.\,\frac{\partial}{\partial z}V_3(x, y, z) + \right.\\ &\qquad\left. + \text{ higher order terms}\right]\mathrm{d}x\,\mathrm{d}y.\end{aligned}$$

So to the first order in $\mathrm{d}z$,

$$\delta F_z = \frac{\partial}{\partial z}\{V_3(x, y, z)\}\,\mathrm{d}z\,\mathrm{d}x\,\mathrm{d}y. \qquad (7)$$

An exactly similar argument for the remaining two pairs of faces gives

$$\left.\begin{aligned} \delta F_x &= \frac{\partial}{\partial x}\{V_1(x,y,z)\}\,\mathrm{d}x\,\mathrm{d}y\,\mathrm{d}z, \\ \delta F_y &= \frac{\partial}{\partial y}\{V_2(x,y,z)\}\,\mathrm{d}x\,\mathrm{d}y\,\mathrm{d}z. \end{aligned}\right\} \tag{8}$$

Adding (7) and (8), we find to the third order in small quantities,

$$\delta F = \left(\frac{\partial V_1}{\partial x} + \frac{\partial V_2}{\partial y} + \frac{\partial V_3}{\partial z}\right)\mathrm{d}x\,\mathrm{d}y\,\mathrm{d}z.$$

Therefore
$$\operatorname{div}\mathbf{V} = \lim_{\delta v \to 0} \frac{\delta F}{\delta v}$$
$$= \frac{\partial V_1}{\partial x} + \frac{\partial V_2}{\partial y} + \frac{\partial V_3}{\partial z}.$$

For any given function $\mathbf{V}$, $\operatorname{div}\mathbf{V}$ is easily found for it is simply a matter of partial differentiation. For example, if $\mathbf{V} = r^2\mathbf{r}$, where $\mathbf{r} = x\mathbf{i} + y\mathbf{j} + z\mathbf{k}$ as usual, then we have

$$V_1 = x(x^2+y^2+z^2), \quad V_2 = y(x^2+y^2+z^2),$$
$$V_3 = z(x^2+y^2+z^2);$$

thus
$$\frac{\partial \mathbf{V}_1}{\partial x} = (x^2+y^2+z^2) + 2x^2,$$

with two similar expressions for $\dfrac{\partial V_2}{\partial y}, \dfrac{\partial V_3}{\partial z}$.

Therefore
$$\operatorname{div}\mathbf{V} = \frac{\partial V_1}{\partial x} + \frac{\partial V_2}{\partial y} + \frac{\partial V_3}{\partial z}$$
$$= 3(x^2+y^2+z^2) + 2x^2 + 2y^2 + 2z^2.$$

Thus
$$\operatorname{div}\mathbf{V} = 5r^2.$$

[A shorter method for this example is given later in 7:8(6).]

Exercises 6:7

1. Evaluate (i) div $(x^3y\mathbf{i} + y^3z\mathbf{j} + z^2\mathbf{k})$,
   (ii) div $[(x^2 + y^2 + z^2)(x\mathbf{i} + y\mathbf{j} - \mathbf{k})]$.
2. Find the divergence of each of the following vector fields: (i) $f(r)\mathbf{r}$, (ii) $(\mathbf{a} \cdot \mathbf{r})\mathbf{r}$, (iii) $r^n f(1/r)\mathbf{r}$, (iv) $(x\mathbf{i} + y\mathbf{j}) \times \mathbf{r}$, (v) $\mathbf{a} \times \mathbf{r}$, (vi) $(x\mathbf{i} + y\mathbf{j}) \times (z\mathbf{j} + x\mathbf{k})$, where $f(1/r)$ is an arbitrary function of $1/r$, $\mathbf{a}$ is a constant vector and n is a constant.

6:8 Expression of the divergence by means of ∇

A most convenient brief expression for div$\mathbf{V}$ is obtained by using the vector operator ∇. Using the normal rule for the scalar multiplication of two vectors, we find that

$$\operatorname{div}\mathbf{V} = \nabla \cdot \mathbf{V}. \tag{1}$$

For
$$\nabla \cdot \mathbf{V} = \left(\mathbf{i}\frac{\partial}{\partial x} + \mathbf{j}\frac{\partial}{\partial y} + \mathbf{k}\frac{\partial}{\partial z}\right) \cdot (V_1\mathbf{i} + V_2\mathbf{j} + V_3\mathbf{k})$$

$$= \frac{\partial}{\partial x}V_1 + \frac{\partial}{\partial y}V_2 + \frac{\partial}{\partial z}V_3\,.$$

Some writers *define* div$\mathbf{V}$ as the result of the scalar multiplication of ∇ and $\mathbf{V}$. It should be stressed that ∇ is a *vector operator*, not a vector, but in combination with vectors it obeys the normal rules of multiplication for vectors. We shall return to this point later when considering identities involving ∇ in § 7:6.

If $\mathbf{V}_a$ and $\mathbf{V}_b$ are two vector point functions, then $\mathbf{V}_a + \mathbf{V}_b$ is also a vector point function, and the normal rule for scalar multiplication of vectors gives at once

$$\operatorname{div}(\mathbf{V}_a + \mathbf{V}_b) = \nabla \cdot (\mathbf{V}_a + \mathbf{V}_b) = \nabla \cdot \mathbf{V}_a + \nabla \cdot \mathbf{V}_b. \tag{2}$$

This result can be extended to the sum of any number of vector point functions.

A very useful identity is

$$\operatorname{div}(\phi\mathbf{V}) = \nabla \cdot (\phi\mathbf{V}) = \phi(\nabla \cdot \mathbf{V}) + \mathbf{V} \cdot (\nabla\phi), \tag{3}$$

where ϕ is a scalar point function. To prove this, we have

$$\nabla . (\phi \mathbf{V}) = \left(\mathbf{i}\frac{\partial}{\partial x} + \mathbf{j}\frac{\partial}{\partial y} + \mathbf{k}\frac{\partial}{\partial z}\right) . (\phi V_1 \mathbf{i} + \phi V_2 \mathbf{j} + \phi V_3 \mathbf{k})$$

$$= \frac{\partial}{\partial x}(\phi V_1) + \frac{\partial}{\partial y}(\phi V_2) + \frac{\partial}{\partial z}(\phi V_3)$$

$$= \phi\left(\frac{\partial V_1}{\partial x} + \frac{\partial V_2}{\partial y} + \frac{\partial V_3}{\partial z}\right) + \left(V_1\frac{\partial \phi}{\partial x} + V_2\frac{\partial \phi}{\partial y} + V_3\frac{\partial \phi}{\partial z}\right).$$

Therefore $$\nabla . (\phi \mathbf{V}) = \phi(\nabla . \mathbf{V}) + \mathbf{V} . (\nabla \phi),$$

since

$$\nabla \phi = \frac{\partial \phi}{\partial x}\mathbf{i} + \frac{\partial \phi}{\partial y}\mathbf{j} + \frac{\partial \phi}{\partial z}\mathbf{k}.$$

This identity is sometimes written more fully in words as

$$\operatorname{div}(\phi \mathbf{V}) = \phi \operatorname{div} \mathbf{V} + \mathbf{V} . \operatorname{grad} \phi.$$

As an example, we can use (3) to find $\operatorname{div}(r^2\mathbf{r})$. The ϕ of (3) is r^2 $(= x^2 + y^2 + z^2)$, and so

$$\nabla . (r^2 \mathbf{r}) = \mathbf{r} . (\nabla r^2) + r^2(\nabla . \mathbf{r}). \tag{4}$$

Now $$\nabla r^2 = \left(\mathbf{i}\frac{\partial}{\partial x} + \mathbf{j}\frac{\partial}{\partial y} + \frac{\partial}{\partial z}\right)\left(x^2 + y^2 + z^2\right)$$

$$= 2x\mathbf{i} + 2y\mathbf{j} + 2z\mathbf{k}.$$

Therefore $$\nabla r^2 = 2\mathbf{r}.$$

So in (4)

$$\nabla . (r^2 \mathbf{r}) = \mathbf{r} . 2\mathbf{r} + r^2\left(\mathbf{i}\frac{\partial}{\partial x} + \mathbf{j}\frac{\partial}{\partial y} + \mathbf{k}\frac{\partial}{\partial z}\right) . (x\mathbf{i} + y\mathbf{j} + z\mathbf{k})$$

$$= 2r^2 + r^2 . 3$$

$$= 5r^2.$$

[This result was obtained by another method in § 6:7; note also the results $\nabla r^2 = 2\mathbf{r}$, $\nabla . \mathbf{r} = 3$ which occur in the above analysis.]

Of especial importance is the expression for div(grad ϕ). Using ∇, this is

$$\nabla \cdot (\nabla\phi) = \left(\mathbf{i}\frac{\partial}{\partial x} + \mathbf{j}\frac{\partial}{\partial y} + \mathbf{k}\frac{\partial}{\partial z}\right) \cdot \left(\frac{\partial\phi}{\partial x}\mathbf{i} + \frac{\partial\phi}{\partial y}\mathbf{j} + \frac{\partial\phi}{\partial z}\mathbf{k}\right)$$

$$= \frac{\partial}{\partial x}\left(\frac{\partial\phi}{\partial x}\right) + \frac{\partial}{\partial y}\left(\frac{\partial\phi}{\partial y}\right) + \frac{\partial}{\partial z}\left(\frac{\partial\phi}{\partial z}\right).$$

Therefore

$$\nabla \cdot (\nabla\phi) = \frac{\partial^2\phi}{\partial x^2} + \frac{\partial^2\phi}{\partial y^2} + \frac{\partial^2\phi}{\partial z^2}. \tag{5}$$

Thus div(grad ϕ), or as it is usually written, div grad ϕ, gives us the combination of second partial derivatives of ϕ which occur in Laplace's equation. This combination is of such importance in so many subjects that we use a special symbol to denote it.

We use ∇^2 to denote the *scalar operator*

$$\frac{\partial^2}{\partial x^2} + \frac{\partial^2}{\partial y^2} + \frac{\partial^2}{\partial z^2}.$$

So

$$\text{div grad } \phi \equiv \nabla^2\phi = \left(\frac{\partial^2\phi}{\partial x^2} + \frac{\partial^2\phi}{\partial y^2} + \frac{\partial^2\phi}{\partial z^2}\right). \tag{6}$$

The scalar operator ∇^2 may also be applied to a vector point function; it then acts on each component of the vector. Thus by definition

$$\nabla^2\mathbf{V} = (\nabla^2 V_1)\mathbf{i} + (\nabla^2 V_2)\mathbf{j} + (\nabla^2 V_3)\mathbf{k}. \tag{7}$$

[For a discussion of the physical significance of $\nabla^2\phi$, the reader is referred to C. G. Lambe and C. J. Tranter, *Differential Equations for Engineers and Scientists*, English Universities Press, 1961, pp. 214–15.]

Exercises 6:8

1. Use the identity 6:8 (3) to find (i) $\nabla \cdot (r^n\mathbf{r})$, (ii) $\nabla \cdot [(\mathbf{a} \cdot \mathbf{r})\mathbf{r}]$, (iii) $\nabla \cdot [r(\nabla r^2)]$, (iv) $\nabla[\nabla \cdot (r^2\mathbf{r})]$, where $\mathbf{a}$ is a constant vector.
2. Using the identity $\nabla^2\phi = \nabla \cdot \nabla\phi$, find (i) $\nabla^2(r^4)$, (ii) $\nabla^2(\log r)$.
3. Show that $\nabla^2(\phi\psi) = \phi\nabla^2\psi + 2\nabla\phi \cdot \nabla\psi + \psi\nabla^2\phi$, where ϕ and ψ are scalar point functions. Hence evaluate $\nabla^2(r^{-1}\log r)$.
4. Evaluate $\nabla^2(r^2\mathbf{r})$.

6:9 Solenoidal vector fields

A vector field **V** which is such that div **V** is zero at *all* points of the field is said to be *solenoidal*. From the definition of the divergence, it follows that if **V** is solenoidal, then the flux through any closed surface in the field is zero. In electrostatics, for example, this is true in a region which has no electric charge. Geometrically, field lines cannot begin or end at points in a solenoidal field. (A *field* line is the graphical representation of a vector point function; it is a line such that the tangent at each point is in the direction of the vector point function **V** at that point.)

EXAMPLE 1. *Show that the vector field* $\mathbf{V} = r^\alpha \mathbf{r}$ *is solenoidal only for* $\alpha = -3$, *but irrotational for all* α.

If **V** is to be solenoidal, then $\nabla . \mathbf{V}$ must be zero; that is

$$\nabla . (r^\alpha \mathbf{r}) = 0.$$

Using 6:8(3), we require

$$r^\alpha(\nabla . \mathbf{r}) + \mathbf{r} . (\nabla r^\alpha) = 0. \tag{1}$$

From the worked example in § 6:8 we have already the result

$$\nabla . \mathbf{r} = 3.$$

Also

$$\nabla r^\alpha = \left(\mathbf{i}\frac{\partial}{\partial x} + \mathbf{j}\frac{\partial}{\partial y} + \mathbf{k}\frac{\partial}{\partial z}\right) r^\alpha$$

$$= \alpha r^{\alpha-1}\left[\frac{\partial r}{\partial x}\mathbf{i} + \frac{\partial r}{\partial y}\mathbf{j} + \frac{\partial r}{\partial z}\mathbf{k}\right]. \tag{2}$$

Since $r^2 = x^2 + y^2 + z^2$, it follows at once by partial differentiation with respect to x, y, and z respectively that

$$r\frac{\partial r}{\partial x} = x, \quad r\frac{\partial r}{\partial y} = y, \quad r\frac{\partial r}{\partial z} = z.$$

So (2) becomes

$$\nabla r^\alpha = \alpha r^{\alpha-2}(x\mathbf{i} + y\mathbf{j} + z\mathbf{k}) = \alpha r^{\alpha-2}\mathbf{r}.$$

Then (1) becomes

$$3r^{\alpha} + \alpha r^{\alpha-2}\mathbf{r} \cdot \mathbf{r} = 0.$$

So

$$3r^{\alpha} + \alpha r^{\alpha-2} \cdot r^2 = 0,$$

and

$$(3 + \alpha)r^{\alpha} = 0.$$

This is true for all r only if $\alpha = -3$.

On the other hand, if $\mathbf{V}$ is irrotational there must exist a function ϕ such that

$$\frac{\partial \phi}{\partial x} = r^{\alpha}x, \quad \frac{\partial \phi}{\partial y} = r^{\alpha}y, \quad \frac{\partial \phi}{\partial z} = r^{\alpha}z. \tag{3}$$

Since

$$\frac{\partial r}{\partial x} = \frac{x}{r}, \quad \frac{\partial r}{\partial y} = \frac{y}{r}, \quad \frac{\partial r}{\partial z} = \frac{z}{r},$$

we can write (3) as

$$\frac{\partial \phi}{\partial x} = r^{\alpha}x = r^{\alpha+1}\frac{\partial r}{\partial x}, \quad \frac{\partial \phi}{\partial y} = r^{\alpha+1}\frac{\partial r}{\partial y}, \quad \frac{\partial \phi}{\partial z} = r^{\alpha+1}\frac{\partial r}{\partial z}.$$

It follows that ϕ can be taken as $r^{\alpha+2}/(\alpha + 2)$, provided that $\alpha \neq -2$. So a scalar function ϕ certainly exists for $\alpha \neq -2$. If $\alpha = -2$, then

$$\frac{\partial \phi}{\partial x} = \frac{x}{x^2 + y^2 + x^2},$$

with similar expressions for $\partial\phi/\partial y$, $\partial\phi/\partial z$. Inspection then shows that we can take

$$\phi = \log \sqrt{(x^2 + y^2 + z^2)}.$$

So for all α, a scalar function ϕ exists such that $\nabla\phi = \mathbf{V}$; that is, $\mathbf{V}$ is irrotational.

EXAMPLE 2. *If ϕ is any solution of Laplace's equation, then $\nabla\phi$ is solenoidal.*

We have

$$\nabla \cdot \nabla\phi = \nabla^2\phi = \frac{\partial^2\phi}{\partial x^2} + \frac{\partial^2\phi}{\partial y^2} + \frac{\partial^2\phi}{\partial z^2},$$

by 6:8 (5) and (6).

So if ϕ is a solution of Laplace's equation, $\nabla \cdot \nabla\phi = 0$; that is $\nabla\phi$ is solenoidal.

Exercises 6:9

1. Show that the vector field $\mathbf{V} = (xyz)^m(x^n\mathbf{i} + y^n\mathbf{j} + z^n\mathbf{k})$ is solenoidal only if $m + n = 0$.
2. Show that $(yz\mathbf{i} + zx\mathbf{j} + xy\mathbf{k})$ is both solenoidal and irrotational. Is this also true for the vector field $\mathbf{V}$, where

$$\mathbf{V} = x(y^2 - z^2)\mathbf{i} + y(z^2 + x^2)\mathbf{j} - z(x^2 - y^2)\mathbf{k}\ ?$$

6:10 The Divergence Theorem

Suppose that we are given a vector field $\mathbf{A}$, such that the components of $\mathbf{A}$ and their partial derivatives are continuous. Let S be a closed surface in the field, enclosing a volume V. Then the Divergence Theorem (or Gauss's Theorem) states that

$$\int_V \operatorname{div} \mathbf{A}\, \mathrm{d}v = \int_S \mathbf{A} \cdot \mathrm{d}\mathbf{S}, \tag{1}$$

where the surface integral is necessarily taken over the outside of S, since $\mathrm{d}\mathbf{S}$ is defined as $\mathrm{d}S\hat{\mathbf{n}}$ and $\hat{\mathbf{n}}$ is the outward unit vector normal to S. Indeed, we may write $(\mathbf{A} \cdot \hat{\mathbf{n}})\mathrm{d}S$ for $\mathbf{A} \cdot \mathrm{d}\mathbf{S}$ on the right of (1).

The cartesian form of (1) is obtained if we write $\mathbf{A} = A_1\mathbf{i} + A_2\mathbf{j} + A_3\mathbf{k}$. Then

$$\iiint_V \left(\frac{\partial A_1}{\partial x} + \frac{\partial A_2}{\partial y} + \frac{\partial A_3}{\partial z}\right) \mathrm{d}x\, \mathrm{d}y\, \mathrm{d}z$$

$$= \iint_S (A_1\, \mathrm{d}y\, \mathrm{d}z + A_2\, \mathrm{d}z\, \mathrm{d}x + A_3\, \mathrm{d}x\, \mathrm{d}y). \tag{2}$$

The reader should compare this with the form given in 5:5(1).

We give a proof of this theorem using the integral definition of div $\mathbf{A}$; this can be omitted by readers interested only in applications. If div $\mathbf{A}$ is defined as in § 6:8 by $\nabla \cdot \mathbf{A}$, then the proof of the Divergence Theorem to be used is that of § 5:5.

We recall that div $\mathbf{A}$ is defined as $\lim\limits_{\delta v \to 0} \delta F/\delta v$. We therefore begin our proof by dividing up the volume V into elements of volume by planes parallel to the coordinate planes. Suppose two such elements have a common plane face parallel to the plane $z = 0$. This face will be the *upper* face for one of the elements and the *lower* face for the other. Hence the flux over the upper face of the first element will cancel out with the flux over the lower face of the second element. This will be true *whenever* two elements have a common plane face, whether it be parallel to $z = 0$, $x = 0$ or $y = 0$. It is only where an element has a surface which is part of S itself that any resultant flux is left when we add for all elements of the volume V.

Suppose δv_p is a typical volume element and δF_p the flux corresponding to it. If F is the flux through the whole surface S, we have already seen that $F = \Sigma \delta F_p$. So, using the definition of flux through a surface (§ 6:6), we have

$$\int_S \mathbf{A} \,.\, d\mathbf{S} = \sum \delta F_p. \tag{3}$$

Our definition of div **A** was

$$\operatorname{div} \mathbf{A} = \lim_{\delta v \to 0} \frac{\delta F}{\delta v}.$$

So for the particular element δv_p,

$$\operatorname{div} \mathbf{A}_p = \lim_{\delta v_p \to 0} \frac{\delta F_p}{\delta v_p}, \tag{4}$$

where $\mathbf{A}_p$ is the value taken by the vector **A** at some point P inside the volume δv_p. It follows that we can write

$$\operatorname{div} \mathbf{A}_p = \frac{\delta F_p}{\delta v_p} + \varepsilon_p, \tag{5}$$

where $\varepsilon_p \to 0$ as $\delta v_p \to 0$; for if in (5) we now let $\delta v_p \to 0$, (4) is obtained. Equation (5) is, in fact, simply expressing what we mean when we say that a function tends to a limit. For convenience, write (5) as

$$\operatorname{div} \mathbf{A}_p \delta v_p = \delta F_p + \varepsilon_p \delta v_p. \tag{6}$$

We sum over all elements δv_p to obtain $\Sigma \delta F_p$; using (6),

$$\sum(\operatorname{div} \mathbf{A}_p \delta v_p) = \sum \delta F_p + \sum(\varepsilon_p \delta v_p). \tag{7}$$

In (7), as $\delta v_p \to 0$, $\Sigma \delta F_p \to F$, the flux through the surface S, and $\Sigma(\operatorname{div} \mathbf{A}_p \, \delta v_p) \to \int_V \operatorname{div} \mathbf{A} \, dv$. So we are left with the term $\Sigma(\varepsilon_p \, \delta v_p)$ to consider, and clearly we want to show that this term tends to zero as $\delta v_p \to 0$. Consider the modulus $|\Sigma(\varepsilon_p \, \delta v_p)|$ of this term. Suppose that $\varepsilon_{\max}$ is the numerically greatest value of ε_p for the whole set of δv_p making up the volume V. Then we obtain at once the inequality

$$\left|\sum(\varepsilon_p \delta v_p)\right| \leqq \varepsilon_{max} \sum(\delta v_p). \tag{8}$$

By our definition of ε_p it follows that $\varepsilon_{\max} \to 0$ as $\delta v_p \to 0$. Hence in (8) we have a product on the right-hand side, one factor of which tends to zero, the other to a finite quantity V as $\delta v_p \to 0$. So the product itself must tend to zero. This proves that $|\Sigma(\varepsilon_p \delta v_p)| \to 0$ and it follows at once that $\Sigma(\varepsilon_p \delta v_p) \to 0$. So (7) gives in t e limit the Divergence Theorem,

$$\int_V \operatorname{div} \mathbf{A} \, dv = \int_S \mathbf{A} \, . \, d\mathbf{S} = \int_S (\mathbf{A} \, . \, \hat{\mathbf{n}}) \, dS.$$

The Divergence Theorem yields many important results when particular field vectors are substituted for **A**. Apart from this, it is often easier to evaluate the volume integral obtained by using the theorem rather than a given surface integral. We shall consider some examples.

EXAMPLE 1. *Show that* $\int_S \mathbf{A} \, . \, d\mathbf{S} = \frac{12}{5}\pi R^5$, *where* S *is the sphere of radius* R, *centre the origin, and* $\mathbf{A} = x^3\mathbf{i} + y^3\mathbf{j} + z^3\mathbf{k}$. [B.Sc.(Eng.)]

By the Divergence Theorem

$$\int_S \mathbf{A} \, . \, d\mathbf{S} = \int_V \operatorname{div} \mathbf{A} \, dv,$$

where in this example V is the volume enclosed by the given sphere. Since

$$\mathbf{A} = x^3\mathbf{i} + y^3\mathbf{j} + z^3\mathbf{k},$$

$$\operatorname{div}\mathbf{A} = 3x^2 + 3y^2 + 3z^2.$$

Therefore $$\int_S \mathbf{A}\,.\,d\mathbf{S} = \int_V 3(x^2 + y^2 + z^2)\,dv,$$

or $$\int_S \mathbf{A}\,.\,d\mathbf{S} = \iiint 3(x^2 + y^2 + z^2)\,dx\,dy\,dz. \qquad (9)$$

To evaluate (9) we change from cartesian to spherical polar coordinates, as in the worked example in § 5:5; the result obtained there was $\frac{12}{5}\pi R^5$ in the notation of this example.

EXAMPLE 2. *Evaluate*

$$\int_S \frac{\hat{\mathbf{n}}\,.\,d\mathbf{S}}{p},$$

where S is the ellipsoid $x^2/a^2 + y^2/b^2 + z^2/c^2 = 1$, p is the perpendicular distance from the origin to the tangent plane at the point on the ellipsoid where d**S** *is situated, and* $\hat{\mathbf{n}}$ *is the unit vector in the direction of the outward normal at that point.* [B.Sc.(Eng.)]

Let d**S** be situated at the point $Q(x_1, y_1, z_1)$. Then the tangent plane at Q has equation

$$\frac{xx_1}{a^2} + \frac{yy_1}{b^2} + \frac{zz_1}{c^2} = 1.$$

Since p is the perpendicular distance from the origin to this plane.

$$p = 1/\sqrt{(x_1^2/a^4 + y_1^2/b^4 + z_1^2/c^4)}. \qquad (10)$$

Let $f(x,y,z) = 0$ be the equation of the ellipsoid; then a vector in the direction of the outward normal at Q is given by grad f, evaluated at $Q(x_1, y_1, z_1)$. Here

$$f(x,y,z) = x^2/a^2 + y^2/b^2 + z^2/c^2 - 1.$$

So if we denote the vector by **n**, then

$$\mathbf{n} = \frac{2x_1}{a^2}\mathbf{i} + \frac{2y_1}{b^2}\mathbf{j} + \frac{2z_1}{c^2}\mathbf{k}. \tag{11}$$

The unit vector $\hat{\mathbf{n}}$ given in the problem is, of course, the unit vector corresponding to **n** in (11). So

$$\hat{\mathbf{n}} = \left(\frac{2x_1}{a^2}\mathbf{i} + \frac{2y_1}{b^2}\mathbf{j} + \frac{2z_1}{c^2}\mathbf{k}\right) \Big/ \sqrt{\left(\frac{4x_1^2}{a^4} + \frac{4y_1^2}{b^4} + \frac{4z_1^2}{c^4}\right)}.$$

Therefore $$\hat{\mathbf{n}} = p\left(\frac{x_1}{a^2}\mathbf{i} + \frac{y_1}{b^2}\mathbf{j} + \frac{z_1}{c^2}\mathbf{k}\right),$$

$$\hat{\mathbf{n}} = \tfrac{1}{2}p\mathbf{n},$$

and $$\frac{\hat{\mathbf{n}} \,.\, \mathrm{d}\mathbf{S}}{p} = \tfrac{1}{2}\mathbf{n} \,.\, \mathrm{d}\mathbf{S}.$$

Applying the Divergence Theorem,

$$\int_S \frac{\hat{\mathbf{n}} \,.\, \mathrm{d}\mathbf{S}}{p} = \int_S \tfrac{1}{2}\mathbf{n} \,.\, \mathrm{d}\mathbf{S} = \int_V \operatorname{div} \mathbf{n} \, \mathrm{d}v,$$

where V is the volume contained by the ellipsoid.

We can now drop the suffixes in **n**, and write

$$\mathbf{n} = \frac{2x}{a^2}\mathbf{i} + \frac{2y}{b^2}\mathbf{j} + \frac{2z}{c^2}\mathbf{k}.$$

So $$\operatorname{div} \mathbf{n} = 2(1/a^2 + 1/b^2 + 1/c^2).$$

Hence $$\int_S \frac{\hat{\mathbf{n}} \,.\, \mathrm{d}\mathbf{S}}{p} = \int_V (1/a^2 + 1/b^2 + 1/c^2)\,\mathrm{d}v$$

$$= (1/a^2 + 1/b^2 + 1/c^2) \,.\, \tfrac{4}{3}\pi abc,$$

since $\int_V \mathrm{d}v$ gives the volume of the ellipsoid, which is $\frac{4}{3}\pi abc$.

EXAMPLE 3. *Evaluate* $\int_S \mathbf{r} \,.\, \hat{\mathbf{n}}\,\mathrm{d}S$, *where* S *is a closed surface,* $\hat{\mathbf{n}}$ *is the unit vector normal to the surface and* **r** *is the position vector of the point where* $\mathrm{d}S$ *is situated.*

Since $d\mathbf{S} = dS\,\hat{\mathbf{n}}$, the Divergence Theorem here gives

$$\int_S \mathbf{r} \cdot \hat{\mathbf{n}}\, dS = \int_S \mathbf{r} \cdot d\mathbf{S} = \int_V \operatorname{div} \mathbf{r}\, dv,$$

where V is the volume enclosed by S.

But since $\mathbf{r} = x\mathbf{i} + y\mathbf{j} + z\mathbf{k}$, $\operatorname{div} \mathbf{r} = 3$. Therefore

$$\int_S \mathbf{r} \cdot \hat{\mathbf{n}}\, dS = \int_V 3\, dv = 3V.$$

We note that we now have an expression for the volume contained by a given closed surface in terms of a surface integral.

EXAMPLE 4. *By taking* $\mathbf{A}$ *in the Divergence Theorem as* $\phi\nabla\psi$, *where* ϕ, ψ *are scalar point functions, establish Green's first identity,*

$$\int_V (\phi\nabla^2\psi + \nabla\phi \cdot \nabla\psi)\, dv = \int_S (\phi\nabla\psi) \cdot d\mathbf{S},$$

and deduce Green's second identity,

$$\int_V (\phi\nabla^2\psi - \psi\nabla^2\phi)\, dv = \int_S (\phi\nabla\psi - \psi\nabla\phi) \cdot d\mathbf{S}.$$

Taking $\mathbf{A} = \phi\nabla\psi$, the Divergence Theorem gives

$$\int_V \operatorname{div}(\phi\nabla\psi)\, dv = \int_S (\phi\nabla\psi) \cdot d\mathbf{S}.$$

But
$$\operatorname{div}(\phi\nabla\psi) = \nabla \cdot (\phi\nabla\psi)$$
$$= \phi\nabla \cdot \nabla\psi + \nabla\phi \cdot \nabla\psi, \text{ by 6:8(3)},$$

so
$$\operatorname{div}(\phi\nabla\psi) = \phi\nabla^2\psi + \nabla\phi \cdot \nabla\psi.$$

Hence
$$\int_V (\phi\nabla^2\psi + \nabla\phi \cdot \nabla\psi)\, dv = \int_S (\phi\nabla\psi) \cdot d\mathbf{S}. \tag{12}$$

Now take $\mathbf{A} = \psi\nabla\phi$. Interchanging ϕ and ψ in (12),

$$\int_V (\psi\nabla^2\phi + \nabla\psi \cdot \nabla\phi)\, dv = \int_S (\psi\nabla\phi) \cdot d\mathbf{S}. \tag{13}$$

From (12) and (13),

$$\int_V (\phi\nabla^2\psi - \psi\nabla^2\phi)\,dv = \int_S (\phi\nabla\psi - \psi\nabla\phi)\,.\,d\mathbf{S}.$$

Exercises 6:10

1. Use the Divergence Theorem to evaluate $\int_S \mathbf{A}\,.\,d\mathbf{S}$, where S is the whole surface of the finite solid bounded by the paraboloid $x^2 + y^2 = z$ and the plane $z = 9$, and $\mathbf{A}$ is given by $\mathbf{A} = (3x + y)\mathbf{i} - x\mathbf{j} + (y - z)\mathbf{k}$. Also find by direct integration the value of $\int_S \mathbf{A}\,.\,d\mathbf{S}$ taken over the plane face of the solid, and hence deduce the value of $\int_S \mathbf{A}\,.\,d\mathbf{S}$ taken over the curved surface only.
2. S is the surface of the solid bounded by the cylinder $x^2 + y^2 = 25$ and the planes $x = 0$, $y = 0$, $z = 0$, $z = 4$. Find the value of $\int_S \mathbf{A}\,.\,d\mathbf{S}$ when (i) $\mathbf{A} = x\mathbf{i} + 3y\mathbf{j} + 4z\mathbf{k}$, (ii) $\mathbf{A} = xy\mathbf{i} + yz\mathbf{j} + zx\mathbf{k}$.
3. Using the notation of worked Example 2 above, show that

$$\operatorname{div}\left(\frac{1}{2p^2}\mathbf{n}\right) = \left(\frac{1}{a^2}+\frac{1}{b^2}+\frac{1}{c^2}\right)\left(\frac{x^2}{a^4}+\frac{y^2}{b^4}+\frac{z^2}{c^4}\right) + 2\left(\frac{x^2}{a^6}+\frac{y^2}{b^6}+\frac{z^2}{c^6}\right).$$

   Hence find the value of $\int_S p^{-3}\,dS$.
4. Verify Green's second identity by evaluating both the volume and surface integrals in the case in which $\phi = x^2y$, $\psi = z^3$, and V is the volume enclosed by the planes $x = 0$, $x = a$, $y = 0$, $y = b$, $z = 0$, $z = c$.
5. Show that

$$\int_S \phi\mathbf{P}\,.\,d\mathbf{S} = \int_V \mathbf{P}\,.\,\operatorname{grad}\phi\,dv + \int_V \phi\operatorname{div}\mathbf{P}\,dv,$$

   where ϕ is a scalar point function, $\mathbf{P}$ a vector point function, and V a volume bounded by a closed surface S.

Miscellaneous Exercises 6

(Throughout these examples, the position vector $x\mathbf{i} + y\mathbf{j} + z\mathbf{k}$ of the point (x, y, z) is denoted by $\mathbf{r}$.)

1. The vector $\mathbf{v}$ and the scalar ϕ are functions of position. Show that (i) the component of $\mathbf{v}$ normal to the level surface ϕ = constant is

$$(\mathbf{v}\,.\,\nabla\phi)\nabla\phi/(\nabla\phi)^2,$$

   (ii) the component of $\mathbf{v}$ tangential to ϕ = constant in the plane of $\mathbf{v}$ and $\hat{\mathbf{n}}$ is

$$\nabla\phi \times (\mathbf{v} \times \nabla\phi)/(\nabla\phi)^2.$$

   [B.Sc.(Eng.)]

2. If $\mathbf{A} = (2xy + 3z^2)\mathbf{i} + (x^2 + 4yz)\mathbf{j} + (2y^2 + 6xz)\mathbf{k}$, evaluate $\int_C \mathbf{A} \cdot d\mathbf{r}$, where C is the curve
   (i) with parametric equations $x = t$, $y = t^2$, $z = t^3$,
   (ii) given by $x = y = z$,
   in each case between the points (0, 0, 0) and (1, 1, 1).
3. Show that the vector field

$$(y^2z^3 \cos x - 4x^3z)\mathbf{i} + (2z^3y \sin x)\mathbf{j} + (3y^2z^2 \sin x - x^4)\mathbf{k}$$

   is irrotational and find the scalar potential ϕ. [B.Sc.(Eng.)]
4. Prove that the vector field $\mathbf{A} = (3x^2 - y)\mathbf{i} + (2yz^2 - x)\mathbf{j} + 2y^2z\mathbf{k}$ is irrotational but not solenoidal. Hence, or otherwise, evaluate the scalar line integral $\int \mathbf{A} \cdot d\mathbf{r}$ along the curve $\mathbf{r} = x\mathbf{i} + (x^2 + x)\mathbf{j} + (2x^3 + x^2)\mathbf{k}$ from (0, 0, 0) to (1, 2, 3).
5. If $\phi = (\mathbf{r} \cdot \mathbf{k})/r^3$, show that

$$\text{grad } \phi = [r^2\mathbf{k} - 3(\mathbf{r} \cdot \mathbf{k})\mathbf{r}]/r^5.$$

   Find also div grad ϕ and hence deduce that a solution of the equation $\nabla^2 f = 0$ is given by $f(x, y, z) = (x + y + z)r^{-3}$. [B.Sc.(Eng.)]
6. If $\mathbf{a}$ is a constant vector, show that

$$\mathbf{a} \cdot \nabla[\mathbf{a} \cdot \nabla(1/r)] = 3(\mathbf{r} \cdot \mathbf{a})^2/r^5 - \mathbf{a}^2/r^3.$$

7. If ϕ is a scalar point function and $\mathbf{A}$ a vector point function, show that

$$\nabla^2(\phi\mathbf{A}) = \phi\nabla^2\mathbf{A} + \mathbf{A}\nabla^2\phi + 2(\nabla\phi \cdot \nabla)\mathbf{A},$$

   where $(\nabla\phi \cdot \nabla)$ is the scalar operator

$$\frac{\partial\phi}{\partial x}\frac{\partial}{\partial x} + \frac{\partial\phi}{\partial y}\frac{\partial}{\partial y} + \frac{\partial\phi}{\partial z}\frac{\partial}{\partial z}.$$

8. If $\mathbf{a}$ and $\mathbf{b}$ are constant vectors and $\phi = \mathbf{a} \cdot \nabla r^3$, $\psi = \mathbf{b} \cdot \nabla r^{-3}$, evaluate
   (i) $\mathbf{b} \cdot \nabla\phi + r^6\mathbf{a} \cdot \nabla\psi$,
   (ii) div $(\phi + \psi)\mathbf{r} - 3(\phi + \psi)$,
   (iii) $\mathbf{b} \cdot$ grad div $(\phi\mathbf{r}) + \mathbf{a} \cdot$ grad div $(r^6\psi\mathbf{r})$.
9. A function $W(x, y)$ satisfies the equation

$$\frac{\partial^2 W}{\partial x^2} + \frac{\partial^2 W}{\partial y^2} + \frac{\partial W}{\partial x} \cdot \frac{\partial V}{\partial x} + \frac{\partial W}{\partial y} \cdot \frac{\partial V}{\partial y} = 0,$$

   where $V(x, y)$ is a given function, such that

$$\frac{\partial^2 V}{\partial x^2} + \frac{\partial^2 V}{\partial y^2} = 0.$$

   By writing the first equation in vector form, show that it has a particular solution $W_0(x, y)$ defined by

$$W_0(x, y) \operatorname{grad} V(x, y) + \operatorname{grad} W_0(x, y) = 0.$$

Hence show that $W_0(x, y) = c\,e^{-V(x,y)}$, where c is a constant.
Show also that the general solution is $W_0\phi$, where ϕ satisfies the equation

$$\frac{\partial}{\partial x}\left(W_0\frac{\partial\phi}{\partial x}\right) + \frac{\partial}{\partial y}\left(W_0\frac{\partial\phi}{\partial y}\right) = 0. \qquad \text{[B.Sc.(Spec.)]}$$

10. A solid of unit density is bounded by a simple closed surface S. The moments of inertia about the axes Ox, Oy, Oz at a point of the solid are A, B, C respectively. If $\mathbf{r}$ is the position vector of any point P on the surface of the solid with respect to the origin O, show that

$$\int_S r^2\mathbf{r}\,.\,\hat{\mathbf{n}}\,\mathrm{d}S = \tfrac{5}{2}(A + B + C),$$

where $\hat{\mathbf{n}}$ is the unit outward vector normal to S at P. [B.Sc.(Gen.)]

11. Find the value of $\int_S \nabla\phi\,.\,\mathrm{d}\mathbf{S}$, where S is the curved surface of the ellipsoidal cap defined by $x^2/a^2 + y^2/b^2 + z^2/c^2 = 1$, $z \geqq 0$, and $\phi = (x+1)^2 + 2(y-1)^2 + z^2$. [Use the Divergence Theorem, and then evaluate $\int \nabla\phi\,.\,\mathrm{d}\mathbf{S}$ over the plane face $z = 0$ of the half-ellipsoid.] [B.Sc.(Eng.)]

12. Starting with the Divergence Theorem $\int_V \operatorname{div}\mathbf{A}\,\mathrm{d}v = \int_S \mathbf{A}\,.\,\mathrm{d}\mathbf{S}$, show by writing $\mathbf{A} = \phi\mathbf{m}$, where $\mathbf{m}$ is a constant vector and ϕ a scalar function of position, that $\mathbf{m}\,.\int_V \nabla\phi\,\mathrm{d}v = \mathbf{m}\,.\int_S \phi\,\mathrm{d}\mathbf{S}$.

Deduce that
$$\int_V \nabla\phi\,\mathrm{d}v = \int_S \phi\,\mathrm{d}\mathbf{S}.$$

Verify this result when $\phi = 3(x^2 + y^2) + z$, and V is the finite part of the paraboloid $x^2 + y^2 = 4 - z$ cut off by the plane $z = 0$.

[On the curved surface of the paraboloid $\mathrm{d}\mathbf{S} = \mathrm{d}S\hat{\mathbf{n}}$, where

$$\mathrm{d}S = \sqrt{(1 + 4x^2 + 4y^2)}\,\mathrm{d}x\,\mathrm{d}y,\ \hat{\mathbf{n}} = (2x\mathbf{i} + 2y\mathbf{j} + \mathbf{k})/\sqrt{(1 + 4x^2 + 4y^2)};$$

integration is over the circle $x^2 + y^2 = 4$.]

13. S is any simple closed surface containing a volume V, $\hat{\mathbf{n}}$ is the unit outward normal to S, and $\mathbf{A} = ax\mathbf{i} + by\mathbf{j} + cz\mathbf{k}$, where a, b, c are constants. Show that
   (i) $\int_V \operatorname{div}\hat{\mathbf{n}}\,\mathrm{d}v = S$,
   (ii) $\int_S (\mathbf{A}\,.\,\hat{\mathbf{n}})\,\mathrm{d}S = (a + b + c)V$,
   (iii) $\int_V (1/r^2)\,\mathrm{d}v = \int_S [(\mathbf{r}\,.\,\hat{\mathbf{n}})/r^2]\,\mathrm{d}S$.

14. A volume V in the positive octant is bounded by the coordinate planes and the sphere $x^2 + y^2 + z^2 = a^2$. If $\mathbf{A} = xz\mathbf{i} + yz\mathbf{j} + z^2\mathbf{k}$, show that $\int_S \mathbf{A}.\mathrm{d}\mathbf{S}$ is zero for each of the three plane faces of V. Hence, using the

Divergence Theorem, show that the value of $\int_S \mathbf{A} \,.\, \mathrm{d}\mathbf{S}$ is $\frac{1}{4}\pi a^4$, where S is the curved surface of V.

Deduce the position of the centroid of the volume V. [B.Sc.(Eng.)]

15. The scalar function $V(x, y, z)$ satisfies Laplace's equation. Show that

$$\int_\tau \left\{\left(\frac{\partial V}{\partial x}\right)^2 + \left(\frac{\partial V}{\partial y}\right)^2 + \left(\frac{\partial V}{\partial z}\right)^2\right\} \mathrm{d}v = \int_S V \frac{\partial V}{\partial n} \mathrm{d}S,$$

$\partial/\partial n$ denoting differentiation along the outward normal to a closed surface S containing a volume τ.

Hence show that if S is the sphere $x^2 + y^2 + z^2 = a^2$, and if V satisfies also the equation $\mathbf{r} \,.\, \nabla V = mV$, where m is a constant, then

$$\int_\tau \left\{\left(\frac{\partial V}{\partial x}\right)^2 + \left(\frac{\partial V}{\partial y}\right)^2 + \left(\frac{\partial V}{\partial z}\right)^2\right\} \mathrm{d}v = \frac{m}{a} \int_S V^2 \,\mathrm{d}S.$$

[V is homogeneous of degree m if $\mathbf{r} \,.\, \nabla V = mV$.]

16. The function $V(x, y)$ satisfies the equations

$$\nabla^2 V = 0, \quad \mathbf{r} \,.\, \nabla V = mV,$$

where m is a constant. Show that

$$\nabla^2(r^{-n}V) + n(2m - n)r^{-n-2}V = 0$$

for any constant n. Hence deduce an expression for

$$\left(\frac{\partial^2}{\partial x^2} + \frac{\partial^2}{\partial y^2}\right)^3 \left(r^{-n}V\right).$$

17. A volume τ is bounded externally by a sphere S with centre C, and internally by a sphere S' with centre C', this second sphere lying wholly within S. Show that

$$\int_S \left(U\frac{\partial V}{\partial n} - V\frac{\partial U}{\partial n}\right) \mathrm{d}S + \int_{S'} \left(U\frac{\partial V}{\partial n} - V\frac{\partial U}{\partial n}\right) \mathrm{d}S = 0,$$

where U, V are scalar functions of position such that $\nabla^2 U = 0$, $\nabla^2 V = 0$ at all points of τ, and $\partial/\partial n$ denotes differentiation along the normal to S or S' in a direction outwards from τ.

18. (i) Show that

$$\int_S \frac{\partial}{\partial n}\left(\frac{1}{r}\right) \mathrm{d}S = 0,$$

where $\mathbf{r} = (x - a)\mathbf{i} + (y - b)\mathbf{j} + (z - c)\mathbf{k}$, S is any simple closed surface, $\partial/\partial n$ denotes differentiation along the outward normal to S and the point (a, b, c) is outside S.

(ii) If S' is the sphere with centre (a, b, c) and radius ρ, show that

$$\int_{S'} \frac{\partial}{\partial n}\left(\frac{1}{r}\right) dS = -\frac{1}{\rho^2}\int_{S'} dS,$$

and deduce the value of the integral on the left as $\rho \to 0$.

Hence show that

$$\int_S \frac{\partial}{\partial n}\left(\frac{1}{r}\right) dS = -4\pi$$

when the point (a, b, c) lies within S.

CHAPTER 7

VECTOR FIELD THEORY (2)

7:1 The curl of a vector

We derived the divergence of a vector from a limit involving flux; we now introduce the *curl* of a vector, which is derived from a limit involving *circulation.* Whereas the divergence is a scalar, the *curl* of a vector is itself a vector.

At a point P in a vector field $\mathbf{V}$, draw a unit vector $\hat{\mathbf{n}}$ in any chosen direction. In the plane through P perpendicular to $\hat{\mathbf{n}}$,

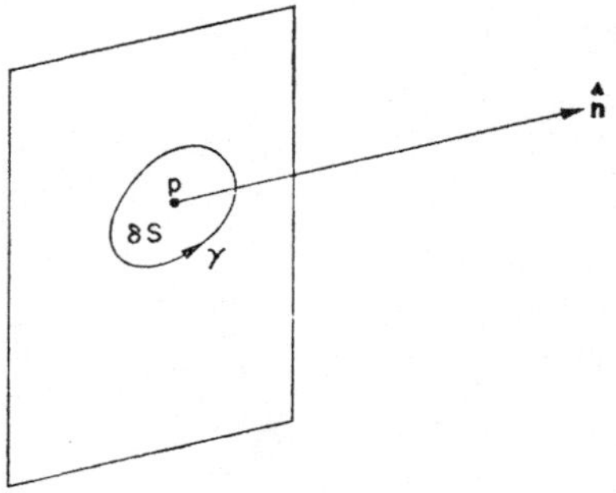

FIG. 34

draw a small closed curve γ round P, enclosing an area δS (see Fig. 34). Denote the circulation of $\mathbf{V}$ in the circuit γ by δC. Then

$$\delta C = \oint_{\gamma} \mathbf{V} \cdot \mathrm{d},$$

from § 6:4.

If now we form the ratio $\delta C/\delta S$ and let δS tend to zero, the limiting quantity we get is a scalar, corresponding to the particular choice of $\hat{\mathbf{n}}$. We define curl **V** as that vector whose component in the direction of $\hat{\mathbf{n}}$ is this limit. So

$$(\text{curl}\,\mathbf{V})_n = \lim_{\delta S \to 0} \frac{\delta C}{\delta S}. \tag{1}$$

Assuming that the vector so defined is unique, we shall establish the analytical form for the components in the directions of **i**, **j**, **k**. Readers interested only in applications may omit this, and accept instead the alternative definition of curl **V** in § 7:3. We show that

$$\text{curl}\,\mathbf{V} = \left(\frac{\partial V_3}{\partial y} - \frac{\partial V_2}{\partial z}\right)\mathbf{i} + \left(\frac{\partial V_1}{\partial z} - \frac{\partial V_3}{\partial x}\right)\mathbf{j} + \left(\frac{\partial V_2}{\partial x} - \frac{\partial V_1}{\partial y}\right)\mathbf{k}. \tag{2}$$

We obtain the components of curl **V** by taking $\hat{\mathbf{n}}$ in turn as **i**, **j**, **k**. We shall consider in detail the case of $\hat{\mathbf{n}} = \mathbf{k}$. **V** is the field vector at the point (x, y, z). Take axes $Px'y'z'$ at P, parallel to the coordinate axes $Oxyz$. Let γ be a simple small closed curve in the (x', y') plane surrounding P (see Fig. 35). Any point Q on γ

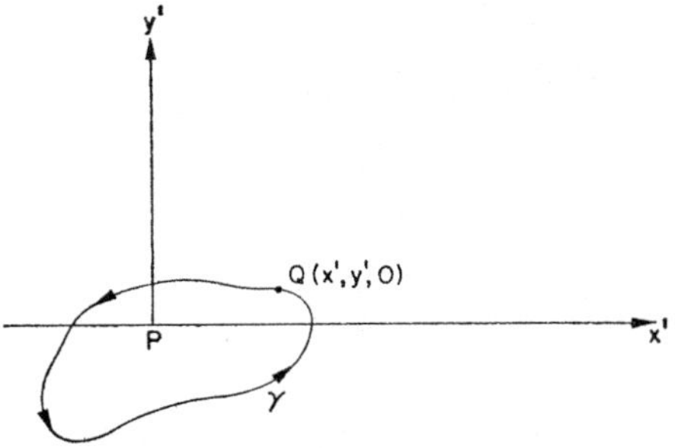

FIG. 35

will have coordinates $(x', y', 0)$ with respect to $Px'y'z'$, and $(x + x', y + y', z)$ with respect to $Oxyz$. Denoting the circulation in the circuit γ by δC_z, we have

$$\delta C_z = \oint_\gamma \mathbf{V} \,.\, d\mathbf{r}, \tag{3}$$

where $\mathbf{r} = (x + x')\mathbf{i} + (y + y')\mathbf{j} + z\mathbf{k}$; as Q moves round γ, only x' and y' change, x, y, z remaining constant. So

$$\mathrm{d}\mathbf{r} = \mathrm{d}x'\mathbf{i} + \mathrm{d}y'\mathbf{j}. \tag{4}$$

Substituting from (4) into (3), and inserting the components of $\mathbf{V}$,

$$\delta C_z = \oint_\gamma [V_1(x + x', y + y', z)\,\mathrm{d}x' + V_2(x + x', y + y', z)\,\mathrm{d}y'].$$

Using Taylor's Theorem, we can expand V_1 and V_2; we may ignore terms involving second and higher powers of $\mathrm{d}x'$ and $\mathrm{d}y'$. Then

$$\delta C_z = \oint_\gamma \left[\left(V_1 + x'\frac{\partial V_1}{\partial x} + y'\frac{\partial V_1}{\partial y}\right)\mathrm{d}x' + \right.$$
$$\left. + \left(V_2 + x'\frac{\partial V_2}{\partial x} + y'\frac{\partial V_2}{\partial y}\right)\mathrm{d}y'\right]. \tag{5}$$

In (5), V_1, V_2 and the partial derivatives refer to $P(x, y, z)$, so they are constant in the line integral round γ. Hence

$$\delta C_z = V_1 \oint_\gamma \mathrm{d}x' + \frac{\partial V_1}{\partial x}\oint_\gamma x'\,\mathrm{d}x' + \frac{\partial V_1}{\partial y}\oint_\gamma y'\,\mathrm{d}x' + V_2\oint_\gamma \mathrm{d}y' +$$
$$+ \frac{\partial V_2}{\partial x}\oint_\gamma x'\,\mathrm{d}y' + \frac{\partial V_2}{\partial y}\oint_\gamma y'\,\mathrm{d}y'.$$

The line integrals are easily evaluated: for

$$\oint_\gamma \mathrm{d}x' = [x']_\gamma = 0; \text{ so also } \oint_\gamma \mathrm{d}y' = 0.$$

$$\oint_\gamma x'\,\mathrm{d}x' = [(x')^2/2]_\gamma = 0; \text{ so also } \oint_\gamma y'\,\mathrm{d}y' = 0.$$

$$\oint_\gamma y'\,\mathrm{d}x' = -\delta S_z, \text{ where } \delta S_z \text{ is the area enclosed by } \gamma.$$

Similarly, $\oint_\gamma x'\,\mathrm{d}y' = \delta S_z$. [Both these results follow by applying

Green's Theorem to the line integrals; see 4:6(6).] Using the values of the separate line integrals, we obtain

$$\delta C_z = -\frac{\partial V_1}{\partial y}\delta S_z + \frac{\partial V_2}{\partial x}\delta S_z.$$

Thus $$\frac{\delta C_z}{\delta S_z} = \frac{\partial V_2}{\partial x} - \frac{\partial V_1}{\partial y}, \quad \text{to a first approximation.}$$

Letting $\delta S_z \to 0$, we find

$$\lim_{\delta S_z \to 0} \frac{\delta C_z}{\delta S_z} = \left(\frac{\partial V_2}{\partial x} - \frac{\partial V_1}{\partial y}\right).$$

Thus the component of curl **V** in the direction of **k** is

$$\left(\frac{\partial V_2}{\partial x} - \frac{\partial V_1}{\partial y}\right).$$

A similar argument gives the two other components, and so

$$\operatorname{curl} \mathbf{V} = \left(\frac{\partial V_3}{\partial y} - \frac{\partial V_2}{\partial z}\right)\mathbf{i} + \left(\frac{\partial V_1}{\partial z} - \frac{\partial V_3}{\partial x}\right)\mathbf{j} + \left(\frac{\partial V_2}{\partial x} - \frac{\partial V_1}{\partial y}\right)\mathbf{k}. \quad (6)$$

Given **V**, (6) defines a unique vector at every point (x, y, z) of the field. Strictly, it is now necessary to check that the component of curl **V**, as given by (6), in the direction of the unit vector $\hat{\mathbf{n}}$, where $\hat{\mathbf{n}}$ is in any chosen direction, does agree with the component of curl **V** in that direction as defined by (1). We shall not pursue this point here; the reader is referred to M. Abraham and R. Becker, *The Classical Theory of Electricity and Magnetism*, Blackie, London, 1932, for details.

The curl of a vector is, in fact, connected with rotation in the vector field. An example illustrating this is given in § 7:3 (Example 4). The notation "rotation of **V**", or "rot **V**" is found in some texts, particularly those of German or Russian origin.

When **V** is given, curl **V** can be found at once from (6) by writing down the appropriate partial derivatives.

Exercises 7:1

1. Given $\mathbf{V} = x^2y\mathbf{i} + (4xz + y^2)\mathbf{j} + (5z^2 + xy^2)\mathbf{k}$, find curl $\mathbf{V}$, and write down its value at the points (1, 1, 1) and (1, 2, 3).
2. Given $\mathbf{V} = xz^3\mathbf{i} - 4x^2yz\mathbf{j} + 3xz^3\mathbf{k}$, find curl (curl $\mathbf{V}$).
3. If $\phi = (x^2y^2 + y^2z^2 + z^2x^2)$, $\mathbf{V} = xy\mathbf{i} + yz\mathbf{j} + zx\mathbf{k}$, evaluate (i) curl grad ϕ, (ii) div curl $\mathbf{V}$.
4. Evaluate (i) curl $r^3\mathbf{r}$, (ii) curl grad $(1/r)$.

7:2 Expression of curl V by means of ∇

If we perform the *vector multiplication* denoted by $\nabla \times \mathbf{V}$, we find that the resultant vector is curl $\mathbf{V}$. For

$$\nabla \times \mathbf{V} = \left(\mathbf{i}\frac{\partial}{\partial x} + \mathbf{j}\frac{\partial}{\partial y} + \mathbf{k}\frac{\partial}{\partial z}\right) \times (V_1\mathbf{i} + V_2\mathbf{j} + V_3\mathbf{k})$$

$$= \begin{vmatrix} \mathbf{i} & \mathbf{j} & \mathbf{k} \\ \dfrac{\partial}{\partial x} & \dfrac{\partial}{\partial y} & \dfrac{\partial}{\partial z} \\ V_1 & V_2 & V_3 \end{vmatrix}, \quad \text{applying 1:4(4),}$$

$$= \mathbf{i}\left(\frac{\partial V_3}{\partial y} - \frac{\partial V_2}{\partial z}\right) + \mathbf{j}\left(\frac{\partial V_1}{\partial z} - \frac{\partial V_3}{\partial x}\right) + \mathbf{k}\left(\frac{\partial V_2}{\partial x} - \frac{\partial V_1}{\partial y}\right). \quad (1)$$

Some writers define curl$\mathbf{V}$ by (1). The integral definition used in § 7:1 has the advantage that it extends easily to the case of curvilinear coordinates, as we shall see in § 7:7.

The determinant form for $\nabla \times \mathbf{V}$ is easy to remember and apply in working out problems. Two important identities are

$$\nabla \times (\mathbf{V}_A + \mathbf{V}_B) = \nabla \times \mathbf{V}_A + \nabla \times \mathbf{V}_B, \quad (2)$$

$$\nabla \times (\phi\mathbf{V}) = (\nabla\phi) \times \mathbf{V} + \phi(\nabla \times \mathbf{V}), \quad (3)$$

where ϕ is a scalar point function. Identity (2) follows at once when $\mathbf{V}_A$, $\mathbf{V}_B$ are written in component form. To prove (3), we note that

$$\nabla \times (\phi \mathbf{V}) = \begin{vmatrix} \mathbf{i} & \mathbf{j} & \mathbf{k} \\ \dfrac{\partial}{\partial x} & \dfrac{\partial}{\partial y} & \dfrac{\partial}{\partial z} \\ \phi V_1 & \phi V_2 & \phi V_3 \end{vmatrix}.$$

Therefore

$$\begin{aligned}\nabla \times (\phi \mathbf{V}) &= \mathbf{i}\left[\frac{\partial}{\partial y}(\phi V_3) - \frac{\partial}{\partial z}(\phi V_2)\right] + \mathbf{j}\left[\frac{\partial}{\partial z}(\phi V_1) - \frac{\partial}{\partial x}(\phi V_3)\right] + \\ &\qquad + \mathbf{k}\left[\frac{\partial}{\partial x}(\phi V_2) - \frac{\partial}{\partial y}(\phi V_1)\right] \\ &= \phi\left[\left(\frac{\partial V_3}{\partial y} - \frac{\partial V_2}{\partial z}\right)\mathbf{i} + \left(\frac{\partial V_1}{\partial z} - \frac{\partial V_3}{\partial x}\right)\mathbf{j} + \left(\frac{\partial V_2}{\partial x} - \frac{\partial V_1}{\partial y}\right)\mathbf{k}\right] + \\ &\qquad + \left[\left(V_3\frac{\partial \phi}{\partial y} - V_2\frac{\partial \phi}{\partial z}\right)\mathbf{i} + \left(V_1\frac{\partial \phi}{\partial z} - V_3\frac{\partial \phi}{\partial x}\right)\mathbf{j} + \right. \\ &\qquad \left. + \left(V_2\frac{\partial \phi}{\partial x} - V_1\frac{\partial \phi}{\partial y}\right)\mathbf{k}\right]. \end{aligned} \tag{4}$$

But
$$\nabla \phi = \frac{\partial \phi}{\partial x}\mathbf{i} + \frac{\partial \phi}{\partial y}\mathbf{j} + \frac{\partial \phi}{\partial z}\mathbf{k}.$$

So
$$(\nabla \phi) \times \mathbf{V} = \begin{vmatrix} \mathbf{i} & \mathbf{j} & \mathbf{k} \\ \dfrac{\partial \phi}{\partial x} & \dfrac{\partial \phi}{\partial y} & \dfrac{\partial \phi}{\partial z} \\ V_1 & V_2 & V_3 \end{vmatrix},$$

and on expansion we find that this is the second vector in (4). Hence

$$\nabla \times (\phi \mathbf{V}) = \phi(\nabla \times \mathbf{V}) + (\nabla \phi) \times \mathbf{V}.$$

In words,
$$\operatorname{curl}(\phi \mathbf{V}) = \phi \operatorname{curl} \mathbf{V} + \operatorname{grad} \phi \times \mathbf{V}.$$

We give some examples involving curl $\mathbf{V}$.

EXAMPLE 1. *Evaluate* curl $r^2\mathbf{r}$, *using* (3) *above.*

Taking $\phi = r^2$, $\mathbf{V} = \mathbf{r}$, we find that

$$\nabla \times (r^2\mathbf{r}) = r^2(\nabla \times \mathbf{r}) + (\nabla r^2) \times \mathbf{r}.$$

But $$\nabla \times \mathbf{r} = \left(\mathbf{i}\frac{\partial}{\partial x} + \mathbf{j}\frac{\partial}{\partial y} + \mathbf{k}\frac{\partial}{\partial z}\right) \times (x\mathbf{i} + y\mathbf{j} + z\mathbf{k}) = \mathbf{0},$$

and $$\nabla r^2 = 2\mathbf{r} \quad \text{(see § 6:8)},$$

so $$\nabla \times (r^2\mathbf{r}) = 2\mathbf{r} \times \mathbf{r} = \mathbf{0}.$$

Thus the curl of $r^2\mathbf{r}$ is identically zero.

EXAMPLE 2. *Show that for any scalar function* ϕ, curl grad $\phi \equiv \mathbf{0}$.

$$\begin{aligned} \operatorname{curl}\operatorname{grad}\phi &= \nabla \times (\nabla\phi) \\ &= \begin{vmatrix} \mathbf{i} & \mathbf{j} & \mathbf{k} \\ \dfrac{\partial}{\partial x} & \dfrac{\partial}{\partial y} & \dfrac{\partial}{\partial z} \\ \dfrac{\partial\phi}{\partial x} & \dfrac{\partial\phi}{\partial y} & \dfrac{\partial\phi}{\partial z} \end{vmatrix}. \end{aligned}$$

So $$\begin{aligned} \operatorname{curl}\operatorname{grad}\phi &= \mathbf{i}\left(\frac{\partial^2\phi}{\partial y\,\partial z} - \frac{\partial^2\phi}{\partial z\,\partial y}\right) + \mathbf{j}\left(\frac{\partial^2\phi}{\partial z\,\partial x} - \frac{\partial^2\phi}{\partial x\,\partial z}\right) + \\ &\qquad + \mathbf{k}\left(\frac{\partial^2\phi}{\partial x\,\partial y} - \frac{\partial^2\phi}{\partial y\,\partial x}\right) \\ &\equiv \mathbf{0}. \end{aligned}$$

[We assume throughout that the scalar functions obey the continuity conditions necessary for the equality of second order partial derivatives such as

$$\frac{\partial^2\phi}{\partial y\,\partial z}, \quad \frac{\partial^2\phi}{\partial z\,\partial y}.]$$

Example 1 is a special case of the result just proved; for if we take $\phi = \frac{1}{4}r^4$, then grad $\phi = r^2\mathbf{r}$.

EXAMPLE 3. *Show that for any vector function* **V**, div curl **V** $\equiv 0$.

div curl **V**

$$= \nabla . (\nabla \times \mathbf{V})$$

$$= \left(\mathbf{i}\frac{\partial}{\partial x} + \mathbf{j}\frac{\partial}{\partial y} + \mathbf{k}\frac{\partial}{\partial z}\right) \cdot \left[\left(\frac{\partial V_3}{\partial y} - \frac{\partial V_2}{\partial z}\right)\mathbf{i} + \right.$$

$$\left. + \left(\frac{\partial V_1}{\partial z} - \frac{\partial V_3}{\partial x}\right)\mathbf{j} + \left(\frac{\partial V_2}{\partial x} - \frac{\partial V_1}{\partial y}\right)\mathbf{k}\right]$$

$$= \frac{\partial}{\partial x}\left(\frac{\partial V_3}{\partial y} - \frac{\partial V_2}{\partial z}\right) + \frac{\partial}{\partial y}\left(\frac{\partial V_1}{\partial z} - \frac{\partial V_3}{\partial x}\right) + \frac{\partial}{\partial z}\left(\frac{\partial V_2}{\partial x} - \frac{\partial V_1}{\partial y}\right)$$

$$= \frac{\partial^2 V_3}{\partial x\, \partial y} - \frac{\partial^2 V_2}{\partial x\, \partial z} + \frac{\partial^2 V_1}{\partial y\, \partial z} - \frac{\partial^2 V_3}{\partial y\, \partial x} + \frac{\partial^2 V_2}{\partial z\, \partial x} - \frac{\partial^2 V_1}{\partial z\, \partial y}$$

$$\equiv 0.$$

EXAMPLE 4. *A rigid body rotates about a fixed point* $O(0,0,0)$ *with constant angular velocity* **ω**. *If* **v** *is the velocity of the point* $P(x,y,z)$ *in the body, show that* **ω** $= \frac{1}{2}$curl **v**.

For rigid-body motion of the type specified, if $\overrightarrow{OP} = \mathbf{r}$,

$$\mathbf{v} = \boldsymbol{\omega} \times \mathbf{r}.$$

Therefore

$$\text{curl}\, \mathbf{v} = \nabla \times (\boldsymbol{\omega} \times \mathbf{r}).$$

Writing $\boldsymbol{\omega} = \omega_1\mathbf{i} + \omega_2\mathbf{j} + \omega_3\mathbf{k}$, we find that

$$\text{curl}\, \mathbf{v} = \begin{vmatrix} \mathbf{i} & \mathbf{j} & \mathbf{k} \\ \frac{\partial}{\partial x} & \frac{\partial}{\partial y} & \frac{\partial}{\partial z} \\ \omega_2 z - \omega_3 y, & \omega_3 x - \omega_1 z, & \omega_1 y - \omega_2 x \end{vmatrix}$$

$$= \mathbf{i}(\omega_1 + \omega_1) + \mathbf{j}(\omega_2 + \omega_2) + \mathbf{k}(\omega_3 + \omega_3).$$

So

$$\boldsymbol{\omega} = \frac{1}{2}\text{curl}\, \mathbf{v}.$$

This example illustrates the statement made in § 7:2 that the curl of a vector is connected with rotation in the field. We shall see in § 8:2 (v) that if the velocity at any point in a body of moving fluid is $\mathbf{q}$, then when there is no rotational motion curl $\mathbf{q} = \mathbf{0}$.

Exercises 7:2

1. If $\mathbf{a}$ is a constant vector, show that
   (i) $\nabla \times (\mathbf{a} \times \mathbf{r}) = 2\mathbf{a}$,
   (ii) curl $[(\mathbf{a} \,.\, \mathbf{r})\mathbf{r}] = \mathbf{a} \times \mathbf{r}$,
   (iii) $\nabla \times [r^{-n}(\mathbf{a} \times \mathbf{r})] = (2 - n)r^{-n}\mathbf{a} + nr^{-n-2}(\mathbf{a} \,.\, \mathbf{r})\mathbf{r}$.
2. Show that curl $(\phi \text{ grad } \phi) = \mathbf{0}$, where ϕ is any scalar point function. Verify this result when $\phi = x^2 + y^2z + z^3$.
3. If $\mathbf{A} = x^2\mathbf{i} + y^2\mathbf{j} + z^2\mathbf{k}$, $\mathbf{B} = x\mathbf{i} + y\mathbf{j} + z\mathbf{k}$, evaluate
   (i) $(\mathbf{A} \times \nabla) \times \mathbf{B}$,
   (ii) $(\mathbf{B} \times \nabla) \times \mathbf{A}$.
   [$(\mathbf{A} \times \nabla)$ is the vector operator
$$\left(y^2\frac{\partial}{\partial z} - z^2\frac{\partial}{\partial y}\right)\mathbf{i} + \left(z^2\frac{\partial}{\partial x} - x^2\frac{\partial}{\partial z}\right)\mathbf{j} + \left(x^2\frac{\partial}{\partial y} - y^2\frac{\partial}{\partial x}\right)\mathbf{k}.\text{]}$$

7:3 Alternative condition for an irrotational vector field

We have already (§ 6:5) defined an irrotational vector field as one in which the circulation in *any circuit* in the field is zero. We know that if $\mathbf{V}$ is irrotational, then there exists a scalar function ϕ such that $\mathbf{V} = \text{grad}\,\phi$. Now we obtain an alternative condition.

If the circulation in *any* circuit is to be zero, then δC of § 7:1 must be zero for any small circuit; it follows at once from the integral definition of curl$\mathbf{V}$ that in this case curl$\mathbf{V}$ is zero. Thus *if* $\mathbf{V}$ *is irrotational, then* curl$\mathbf{V} = \mathbf{0}$.

Conversely, *if* curl$\mathbf{V} = \mathbf{0}$, *then* $\mathbf{V}$ *is irrotational.* To prove this we notice that if curl$\mathbf{V} = \mathbf{0}$, then

$$\frac{\partial V_3}{\partial y} = \frac{\partial V_2}{\partial z}, \quad \frac{\partial V_1}{\partial z} = \frac{\partial V_3}{\partial x}, \quad \frac{\partial V_2}{\partial x} = \frac{\partial V_1}{\partial y}. \tag{1}$$

We now show that (1) implies that there exists a scalar potential function ϕ such that $\mathbf{V} = \nabla\phi$. This in turn implies that $\mathbf{V}$ is irrotational.

Consider the line integral $\int \mathbf{V} \cdot d\mathbf{r}$ along a path $ABCD$, from $A(x_0, y_0, z_0)$ to $D(x, y, z)$, made up of lines parallel to the axes as in Fig. 36. On AB only x varies, y and z staying constant at

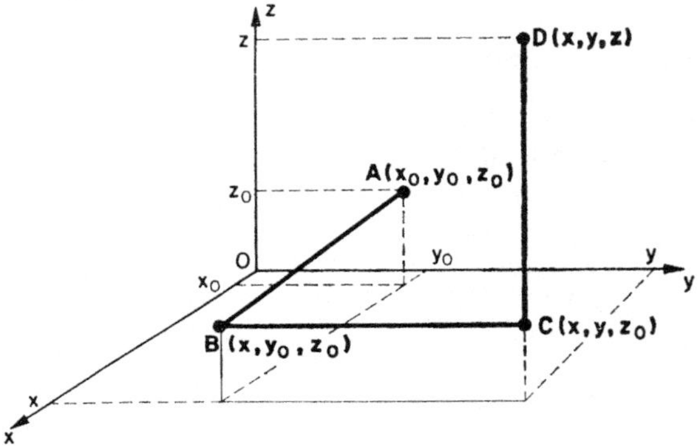

FIG. 36

y_0, z_0 respectively. On BC only y varies, and on CD only z varies. Thus on AB, $dy = dz = 0$, with similar results for the other two parts of the path. Since $\mathbf{V} \cdot d\mathbf{r} = V_1 dx + V_2 dy + V_3 dz$, it follows that

$$\int_{ABCD} \mathbf{V} \cdot d\mathbf{r} = \int_{AB} \mathbf{V} \cdot d\mathbf{r} + \int_{BC} \mathbf{V} \cdot d\mathbf{r} + \int_{CD} \mathbf{V} \cdot d\mathbf{r}$$

$$= \int_{x_0}^{x} V_1(x, y_0, z_0)\, dx + \int_{y_0}^{y} V_2(x, y, z_0)\, dy +$$

$$+ \int_{z_0}^{z} V_3(x, y, z)\, dz. \qquad (2)$$

Denoting the value of $\int_{ABCD} \mathbf{V} \cdot d\mathbf{r}$ by $\phi(x, y, z)$, we then have

$$\frac{\partial \phi}{\partial z} = V_3(x, y, z), \qquad (3)$$

since $z = z_0 =$ constant in the other two integrals.

Also, $$\frac{\partial\phi}{\partial y} = V_2(x, y, z_0) + \int_{z_0}^{z} \frac{\partial V_3}{\partial y} \mathrm{d}z,$$

$$= V_2(x, y, z_0) + \int_{z_0}^{z} \frac{\partial V_2}{\partial z} \mathrm{d}z, \quad \text{by (1).}$$

Thus $$\frac{\partial\phi}{\partial y} = V_2(x, y, z_0) + [V_2(x, y, z)]_{z_0}^{z}$$

$$= V_2(x, y, z_0) + V_2(x, y, z) - V_2(x, y, z_0),$$

i.e. $$\frac{\partial\phi}{\partial y} = V_2(x, y, z). \tag{4}$$

Similarly,

$$\frac{\partial\phi}{\partial x} = V_1(x, y_0, z_0) + \int_{y_0}^{y} \frac{\partial V_2}{\partial x} \mathrm{d}y + \int_{z_0}^{z} \frac{\partial V_3}{\partial x} \mathrm{d}z$$

$$= V_1(x, y_0, z_0) + \int_{y_0}^{y} \frac{\partial V_1}{\partial y} \mathrm{d}y + \int_{z_0}^{z} \frac{\partial V_1}{\partial z} \mathrm{d}z, \quad \text{by (1),}$$

$$= V_1(x, y_0, z_0) + [V_1(x, y, z_0)]_{y_0}^{y} + [V_1(x, y, z)]_{z_0}^{z}$$

$$= V_1(x, y_0, z_0) + V_1(x, y, z_0) - V_1(x, y_0, z_0) +$$
$$+ V_1(x, y, z) - V_1(x, y, z_0);$$

so $$\frac{\partial\phi}{\partial x} = V_1(x, y, z). \tag{5}$$

From (3), (4) and (5),

$$\mathbf{V} = V_1\mathbf{i} + V_2\mathbf{j} + V_3\mathbf{k} = \frac{\partial\phi}{\partial x}\mathbf{i} + \frac{\partial\phi}{\partial y}\mathbf{j} + \frac{\partial\phi}{\partial z}\mathbf{k}.$$

Thus a scalar potential function ϕ exists such that $\mathbf{V} = \nabla\phi$, and hence $\mathbf{V}$ is irrotational.

A vector field $\mathbf{V}$ such that curl $\mathbf{V}$ is *not everywhere zero* is said to be *rotational.*

EXAMPLE. *Show that $r^\alpha \mathbf{r}$ is irrotational for all α.*

Here, $$V_1 = r^\alpha x,\ V_2 = r^\alpha y,\ V_3 = r^\alpha z.$$

So $$\frac{\partial V_3}{\partial y} = \alpha r^{\alpha-1}\, z\frac{\partial r}{\partial y} = \alpha yzr^{\alpha-2},$$

$$\frac{\partial V_2}{\partial z} = \alpha r^{\alpha-1} y\frac{\partial r}{\partial z} = \alpha yzr^{\alpha-2}.$$

Hence $\partial V_3/\partial y = \partial V_2/\partial z$, and similarly

$$\frac{\partial V_1}{\partial z} = \frac{\partial V_3}{\partial x},\quad \frac{\partial V_2}{\partial x} = \frac{\partial V_1}{\partial y}.$$

So curl $\mathbf{V} = \mathbf{0}$, where $\mathbf{V} = r^\alpha \mathbf{r}$. [This example was considered in § 6:9 by another method.]

Exercises 7:3

1. A vector field $\mathbf{V}$ is given by $\mathbf{V} = (xyz)^m(x^n\mathbf{i} + y^n\mathbf{j} + z^n\mathbf{k})$, where m, n are constants. For what values of m and n is the field irrotational? [B.Sc.(Eng.)]

2. Which of the following vector fields are irrotational?
   (i) $r(x^2\mathbf{i} + xy\mathbf{j} + zr\mathbf{k})$,
   (ii) $r\mathbf{a} + r^{-1}(\mathbf{a} \cdot \mathbf{r})\mathbf{r}$,
   (iii) $(\mathbf{a} \times \mathbf{r}) - \nabla \times [(\mathbf{a} \cdot \mathbf{r})\mathbf{r}]$,
   where $\mathbf{a}$ is a constant vector.

7:4 Stokes's Theorem

Suppose that C is a simple closed curve in a vector field $\mathbf{A}$, and that S is an open surface spanning C. If the components of $\mathbf{A}$ and their first partial derivatives are continuous on C and S, then Stokes's Theorem states that

$$\oint_C \mathbf{A} \cdot d\mathbf{r} = \int_S \text{curl}\,\mathbf{A} \cdot d\mathbf{S}, \tag{1}$$

the line integral being taken in the positive sense. (When C is traversed in this sense, the outside of S is to the left; the surface integral is, of course, taken over the outside of S, since $d\mathbf{S} = dS\hat{\mathbf{n}}$ and $\hat{\mathbf{n}}$ is the unit outward vector normal to S.)

In words, Stokes's Theorem states that *the circulation of* **A** *in the circuit C is equal to the flux of* curl**A** *through the surface S.*

The cartesian form of (1) is obtained if we write $\mathbf{A} = A_1\mathbf{i} + A_2\mathbf{j} + A_3\mathbf{k}$. Then (1) becomes

$$\oint_C (A_1\,\mathrm{d}x + A_2\,\mathrm{d}y + A_3\,\mathrm{d}z)$$

$$= \iint_S \left[\left(\frac{\partial A_3}{\partial y} - \frac{\partial A_2}{\partial z}\right)\mathrm{d}y\,\mathrm{d}z + \left(\frac{\partial A_1}{\partial z} - \frac{\partial A_3}{\partial x}\right)\mathrm{d}z\,\mathrm{d}x + \right.$$

$$\left. + \left(\frac{\partial A_2}{\partial x} - \frac{\partial A_1}{\partial y}\right)\mathrm{d}x\,\mathrm{d}y\right].$$

The reader should compare this form with that given in 5:5(5).

We give a proof of this theorem using the integral definition of curl**A**; this can be omitted by readers interested only in applications. If curl**A** is defined as in § 7:3 by $\nabla \times \mathbf{A}$, then the proof of Stokes's Theorem is that of § 5:5.

To prove the result, divide the surface S into elements of surface, δS_p being a typical element surrounding the point P. Consider the circulation round the boundary of each element; if we add these circulations for the whole set, then the result is the circulation round the whole boundary curve C of S, for the line integrals corresponding to internal boundaries occur twice, in opposite directions, and cancel each other out (see Fig. 37).

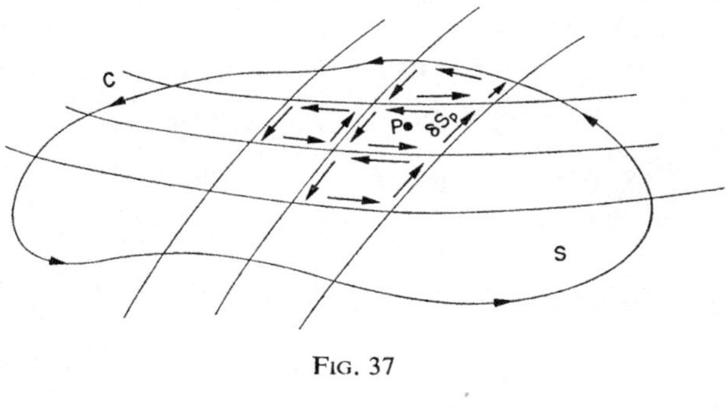

FIG. 37

Now suppose that $\hat{\mathbf{n}}$ is the unit outward vector normal to S at the point P in δS_p. Then

$$(\text{curl } \mathbf{A})_n = \lim_{\delta S_p \to 0} \frac{\delta C_p}{\delta S_p} \tag{1a}$$

in the notation of § 7:1. So we can write

$$(\text{curl } \mathbf{A})_n \delta S_p = \delta C_p + \varepsilon_p \delta S_p,$$

where $\varepsilon_p \to 0$ as $\delta S_p \to 0$.

If we now sum over the whole set of surface elements,

$$\sum [(\text{curl } \mathbf{A})_n \delta S_p] = \sum \delta C_p + \sum (\varepsilon_p \delta S_p). \tag{2}$$

In (2), $\Sigma \delta C_p$ is, as we have seen above, the circulation of $\mathbf{A}$ in the circuit C. So

$$\sum \delta C_p = \oint_C \mathbf{A} \,.\, d\mathbf{r}.$$

Also, since $(\text{curl}\, \mathbf{A})_n = \text{curl}\, \mathbf{A} \,.\, \hat{\mathbf{n}}$, we find that

$$(\text{curl}\, \mathbf{A})_n \delta S_p = \text{curl}\, \mathbf{A} \,.\, \delta \mathbf{S}_p,$$

where $\delta \mathbf{S}_p$ is the vector element of area corresponding to δS_p. So in (2), when $\delta S_p \to 0$,

$$\sum [(\text{curl}\, \mathbf{A})_n \delta S_p] \to \int_S \text{curl}\, \mathbf{A} \,.\, d\mathbf{S}.$$

We are left with the term $\Sigma(\varepsilon_p \delta S_p)$, and we treat this exactly as we did the term $\Sigma(\varepsilon_p \delta v_p)$ in the proof of the Divergence Theorem. If ε_{max} is the numerically greatest value of ε_p, then

$$\left| \sum \varepsilon_p \delta S_p \right| \leqq \varepsilon_{\text{max}} \sum \delta S_p.$$

But $\varepsilon_{\text{max}} \Sigma \delta S_p = \varepsilon_{\text{max}} S$, and from the definition of ε_p, $\varepsilon_{\text{max}} \to 0$ as $\delta S_p \to 0$; so if in (2) we now let $\delta S_p \to 0$ we obtain

$$\int_S \text{curl}\, \mathbf{A} \,.\, d\mathbf{S} = \oint_C \mathbf{A} \,.\, d\mathbf{r}.$$

We give some examples of the use of Stokes's Theorem.

EXAMPLE 1. *Evaluate the integral* $\int_S \text{curl}\,\mathbf{A}\,.\,\text{d}\mathbf{S}$, *where*

$$\mathbf{A} = (x^2 + y - 4)\mathbf{i} + 3xy\mathbf{j} + (2xz + z^2)\mathbf{k},$$

and S is the surface of the sphere $x^2 + y^2 + z^2 = 16$ *for which* $z \geqq 0$. [B.Sc.(Eng.)]

If we apply Stokes's Theorem, we have instead of the given surface integral a line integral round the circle $x^2 + y^2 = 16$, $z = 0$, to evaluate. Now

$$\int_S \text{curl}\,\mathbf{A}\,.\,\text{d}\mathbf{S} = \oint_C \mathbf{A}\,.\,\text{d}\mathbf{r}. \tag{3}$$

In (3), on the circle C, $\text{d}\mathbf{r} = \text{d}x\mathbf{i} + \text{d}y\mathbf{j}$, since $z = 0$; also on this circle, $\mathbf{A} = (x^2 + y - 4)\mathbf{i} + 3xy\mathbf{j}$. Therefore

$$\oint_C \mathbf{A}\,.\,\text{d}\mathbf{r} = \oint_C [(x^2 + y - 4)\,\text{d}x + 3xy\,\text{d}y]. \tag{4}$$

In (4), $\oint_C x^2\,\text{d}x = [\tfrac{1}{3}x^3]_C = 0, \quad \oint_C -4\,\text{d}x = [-4x]_C = 0,$

and $\oint_C y\,\text{d}x = -16\pi$, since by Green's Theorem $-\oint_C y\,\text{d}x$ gives the area D enclosed by C. Again, using Green's Theorem,

$$\oint_C 3xy\,\text{d}y = \iint_D 3y\,\text{d}x\,\text{d}y = 0, \quad \text{by symmetry.}$$

Substituting these results in (4), we find that

$$\int_S \text{curl}\,\mathbf{A}\,.\,\text{d}\mathbf{S} = -16\pi.$$

Note that in this example, as in many examples in practice, C is a plane curve; this is, of course, a special case of the general result, where C need not lie in a plane.

EXAMPLE 2. *Show that* $\int_S \hat{\mathbf{n}}\,.\,\text{curl}\,\mathbf{A}\,\text{d}S = 0$ *when the integral is taken over a closed surface S,* $\hat{\mathbf{n}}$ *being the unit vector normal to the element of surface* dS.

Since $dS\hat{\mathbf{n}} = d\mathbf{S}$, we wish to show that $\int_S \text{curl}\,\mathbf{A}\,.\,d\mathbf{S}$ is zero over a closed surface S. We can divide S into two parts S_1 and S_2 by drawing on the surface any convenient closed curve C. We apply Stokes's Theorem to each part; then, clearly, if the line integral round C corresponding to the part S_1 is taken in anti-clockwise direction to give the correct positive sense, the line integral round C corresponding to the part S_2 of the surface must be taken in the opposite direction. Hence, when we add the two together to obtain the integral over the whole surface, the result is given by

$$\begin{aligned}\int_S \text{curl}\,\mathbf{A}\,.\,d\mathbf{S} &= \int_{S_1} \text{curl}\,\mathbf{A}\,.\,d\mathbf{S} + \int_{S_2} \text{curl}\,\mathbf{A}\,.\,d\mathbf{S} \\ &= \oint_C \mathbf{A}\,.\,d\mathbf{r} + \left(-\oint_C \mathbf{A}\,.\,d\mathbf{r}\right) \\ &= 0.\end{aligned}$$

Exercises 7:4

1. Evaluate $\int \text{curl}\,\mathbf{A}\,.\,d\mathbf{S}$ over that part of the surface $x^2 + 4y^2 + z^2 - 2z = 4$ lying above the plane $z = 0$, given that $\mathbf{A} = (x^3 - y^3)\mathbf{i} - xyz\mathbf{j} + y^3\mathbf{k}$.
2. Given that $\mathbf{A} = 2y(1 - x)\mathbf{i} + (x - x^2 + y^2)\mathbf{j} + (x^2 + y^2 + z^2)\mathbf{k}$, find the value of $\int \hat{\mathbf{n}}\,.\,\text{curl}\,\mathbf{A}\,dS$ taken over the surface of the hemispherical cap for which $x^2 + y^2 + z^2 = 1$, $z \geqq 0$, where $\hat{\mathbf{n}}$ is a unit vector in the direction of the outward normal at the point where dS is situated. [B.Sc.(Spec.)]
3. Starting with Stokes's Theorem

$$\oint_C \mathbf{A}\,.\,d\mathbf{r} = \int_S (\nabla \times \mathbf{A})\,.\,d\mathbf{S},$$

show, by writing $\mathbf{A} = \phi\mathbf{V}$, where ϕ is a scalar point function and $\mathbf{V}$ is an arbitrary constant vector, that

$$\oint_C \phi\,d\mathbf{r} = \int_S (d\mathbf{S} \times \nabla\phi).$$

4. Evaluate the integral $\int \text{curl}\,\mathbf{A}\,.\,d\mathbf{S}$ for the vector function

$$\mathbf{A} = (2y^2 + 3z^2 - x^2)\mathbf{i} + (2z^2 + 3x^2 - y^2)\mathbf{j} + (2x^2 + 3y^2 - z^2)\mathbf{k},$$

over the part of the surface $x^2 + y^2 - 2ax + az = 0$ which lies above the plane $z = 0$. [B.Sc.(Eng.)]

7:5 Summary of the main results for gradient, divergence and curl

It is convenient here to list the more important formulae so far. These are as follows:

$$\operatorname{grad} \phi = \nabla\phi = \frac{\partial \phi}{\partial x}\mathbf{i} + \frac{\partial \phi}{\partial y}\mathbf{j} + \frac{\partial \phi}{\partial z}\mathbf{k}, \tag{1}$$

$$\operatorname{div} \mathbf{V} = \nabla . \mathbf{V} = \frac{\partial V_1}{\partial x} + \frac{\partial V_2}{\partial y} + \frac{\partial V_3}{\partial z}, \tag{2}$$

$$\operatorname{curl} \mathbf{V} = \nabla \times \mathbf{V} = \begin{vmatrix} \mathbf{i} & \mathbf{j} & \mathbf{k} \\ \dfrac{\partial}{\partial x} & \dfrac{\partial}{\partial y} & \dfrac{\partial}{\partial z} \\ V_1 & V_2 & V_3 \end{vmatrix},$$

or

$$\operatorname{curl} \mathbf{V} = \left(\frac{\partial V_3}{\partial y} - \frac{\partial V_2}{\partial z}\right)\mathbf{i} + \left(\frac{\partial V_1}{\partial z} - \frac{\partial V_3}{\partial x}\right)\mathbf{j} + \left(\frac{\partial V_2}{\partial x} - \frac{\partial V_1}{\partial y}\right)\mathbf{k}, \tag{3}$$

$$\nabla^2\phi = \nabla . \nabla\phi = \frac{\partial^2 \phi}{\partial x^2} + \frac{\partial^2 \phi}{\partial y^2} + \frac{\partial^2 \phi}{\partial z^2}, \tag{4}$$

$$\nabla(\phi_1 + \phi_2) = \nabla\phi_1 + \nabla\phi_2, \tag{5}$$

$$\nabla . (\mathbf{V}_A + \mathbf{V}_B) = \nabla . \mathbf{V}_A + \nabla . \mathbf{V}_B, \tag{6}$$

$$\nabla \times (\mathbf{V}_A + \mathbf{V}_B) = \nabla \times \mathbf{V}_A + \nabla \times \mathbf{V}_B, \tag{7}$$

$$\nabla(\phi_1\phi_2) = \phi_1\nabla\phi_2 + \phi_2\nabla\phi_1, \tag{8}$$

$$\nabla(\phi\mathbf{V}) = \phi(\nabla . \mathbf{V}) + \mathbf{V} . \nabla\phi, \tag{9}$$

$$\nabla \times (\phi\mathbf{V}) = (\nabla\phi) \times \mathbf{V} + \phi(\nabla \times \mathbf{V}), \tag{10}$$

$$\text{curl}\,\text{grad}\,\phi = \nabla \times (\nabla\phi) \equiv \mathbf{0}, \tag{11}$$

$$\text{div}\,\text{curl}\,\mathbf{V} = \nabla \cdot (\nabla \times \mathbf{V}) \equiv 0. \tag{12}$$

Many other useful identities can be established; we give the reader some examples below. In examples of this type, it is often possible to shorten the working by the use of (4)–(12) above, though in some cases it is necessary to write out the components in full.

EXAMPLE 1. *Show that*

$$\text{(i)}\quad \nabla \cdot (\mathbf{A} \times \mathbf{B}) = (\nabla \times \mathbf{A}) \cdot \mathbf{B} - \mathbf{A} \cdot (\nabla \times \mathbf{B}), \tag{13}$$

$$\text{(ii)}\quad \nabla \times (\mathbf{A} \times \mathbf{B}) = (\nabla \cdot \mathbf{B})\mathbf{A} - (\nabla \cdot \mathbf{A})\mathbf{B} + (\mathbf{B} \cdot \nabla)\mathbf{A} - (\mathbf{A} \cdot \nabla)\mathbf{B}. \tag{14}$$

$$\begin{aligned}
\text{(i)}\quad \nabla \cdot (\mathbf{A} \times \mathbf{B}) &= \left(\mathbf{i}\frac{\partial}{\partial x} + \mathbf{j}\frac{\partial}{\partial y} + \mathbf{k}\frac{\partial}{\partial z}\right) \cdot [(A_2B_3 - A_3B_2)\mathbf{i} + \\
&\qquad + (A_3B_1 - A_1B_3)\mathbf{j} + (A_1B_2 - A_2B_1)\mathbf{k}] \\
&= \frac{\partial}{\partial x}(A_2B_3 - A_3B_2) + \frac{\partial}{\partial y}(A_3B_1 - A_1B_3) + \\
&\qquad + \frac{\partial}{\partial z}(A_1B_2 - A_2B_1) \\
&= \left(B_3\frac{\partial A_2}{\partial x} - B_2\frac{\partial A_3}{\partial x} + B_1\frac{\partial A_3}{\partial y} - B_3\frac{\partial A_1}{\partial y} + \right. \\
&\qquad \left. + B_2\frac{\partial A_1}{\partial z} - B_1\frac{\partial A_2}{\partial z}\right) \\
&\quad + \left(A_2\frac{\partial B_3}{\partial x} - A_3\frac{\partial B_2}{\partial x} + A_3\frac{\partial B_1}{\partial y} - A_1\frac{\partial B_3}{\partial y} + \right. \\
&\qquad \left. + A_1\frac{\partial B_2}{\partial z} - A_2\frac{\partial B_1}{\partial z}\right)
\end{aligned}$$

$$= \left[B_1\left(\frac{\partial A_3}{\partial y} - \frac{\partial A_2}{\partial z}\right) + B_2\left(\frac{\partial A_1}{\partial z} - \frac{\partial A_3}{\partial x}\right) + \right.$$

$$\left. + B_3\left(\frac{\partial A_2}{\partial x} - \frac{\partial A_1}{\partial y}\right)\right] - \left[A_1\left(\frac{\partial B_3}{\partial y} - \frac{\partial B_2}{\partial z}\right) + \right.$$

$$\left. + A_2\left(\frac{\partial B_1}{\partial z} - \frac{\partial B_3}{\partial x}\right) + A_3\left(\frac{\partial B_2}{\partial x} - \frac{\partial B_1}{\partial y}\right)\right].$$

Therefore

$$\begin{aligned}\nabla . (\mathbf{A} \times \mathbf{B}) &= \mathbf{B} . (\nabla \times \mathbf{A}) - \mathbf{A} . (\nabla \times \mathbf{B}) \\ &= (\nabla \times \mathbf{A}) . \mathbf{B} - \mathbf{A} . (\nabla \times \mathbf{B}).\end{aligned}$$

(ii) In (14), $(\mathbf{B} . \nabla)$ is to be interpreted as an operator; if we write the scalar product in full we get

$$\begin{aligned}\mathbf{B} . \nabla &= (B_1\mathbf{i} + B_2\mathbf{j} + B_3\mathbf{k}) . \left(\mathbf{i}\frac{\partial}{\partial x} + \mathbf{j}\frac{\partial}{\partial y} + \mathbf{k}\frac{\partial}{\partial z}\right) \\ &= \left(B_1\frac{\partial}{\partial x} + B_2\frac{\partial}{\partial y} + B_3\frac{\partial}{\partial z}\right),\end{aligned}$$

which is a scalar operator. Acting on the vector **A**, it will produce a vector; the x-component of this vector, for example, is

$$\left(B_1\frac{\partial A_1}{\partial x} + B_2\frac{\partial A_1}{\partial y} + B_3\frac{\partial A_1}{\partial z}\right),$$

with similar terms for the other two.

We require $\nabla \times (\mathbf{A} \times \mathbf{B})$; so

$$\nabla \times (\mathbf{A} \times \mathbf{B}) = \begin{vmatrix} \mathbf{i} & \mathbf{j} & \mathbf{k} \\ \dfrac{\partial}{\partial x} & \dfrac{\partial}{\partial y} & \dfrac{\partial}{\partial z} \\ A_2B_3 - A_3B_2, & A_3B_1 - A_1B_3, & A_1B_2 - A_2B_1 \end{vmatrix}$$

or $\nabla \times (\mathbf{A} \times \mathbf{B}) = \mathbf{C} = C_1\mathbf{i} + C_2\mathbf{j} + C_3\mathbf{k}$, where

$$C_1 = \left[A_1\frac{\partial B_2}{\partial y} - A_2\frac{\partial B_1}{\partial y} + A_1\frac{\partial B_3}{\partial z} - A_3\frac{\partial B_1}{\partial z}\right] -$$

$$-\left[B_1\frac{\partial A_2}{\partial y} - B_2\frac{\partial A_1}{\partial y} + B_1\frac{\partial A_3}{\partial z} - B_3\frac{\partial A_1}{\partial z}\right].$$

We now group the terms, taking first those with common factors A_1 and B_1; then

$$C_1 = A_1\left(\frac{\partial B_2}{\partial y} + \frac{\partial B_3}{\partial z}\right) - B_1\left(\frac{\partial A_2}{\partial y} + \frac{\partial A_3}{\partial z}\right) +$$

$$+\left(B_2\frac{\partial A_1}{\partial y} + B_3\frac{\partial A_1}{\partial z}\right) - \left(A_2\frac{\partial B_1}{\partial y} + A_3\frac{\partial B_1}{\partial z}\right), \text{ etc.}$$

If we add $\partial B_1/\partial x$ to the first bracket, we have there $\nabla . \mathbf{B}$. Adding $\partial B_1/\partial x$ to the first bracket means adding $A_1(\partial B_1/\partial x)$ to C_1; so we must subtract it again to leave C_1 unaltered. The place for the subtraction is obviously the last bracket. We deal similarly with the second bracket, adding $\partial A_1/\partial x$ to it and subtracting $B_1(\partial A_1/\partial x)$ in the third bracket. Thus we obtain

$$C_1 = A_1\left(\frac{\partial B_1}{\partial x} + \frac{\partial B_2}{\partial y} + \frac{\partial B_3}{\partial z}\right) - B_1\left(\frac{\partial A_1}{\partial x} + \frac{\partial A_2}{\partial y} + \frac{\partial A_3}{\partial z}\right) +$$

$$+\left(B_1\frac{\partial A_1}{\partial x} + B_2\frac{\partial A_1}{\partial y} + B_3\frac{\partial A_1}{\partial z}\right) -$$

$$-\left(A_1\frac{\partial B_1}{\partial x} + A_2\frac{\partial B_1}{\partial y} + A_3\frac{\partial B_1}{\partial z}\right).$$

Therefore

$$C_1 = (\nabla . \mathbf{B})A_1 - (\nabla . \mathbf{A})B_1 + (\mathbf{B} . \nabla)A_1 - (\mathbf{A} . \nabla)B_1.$$

C_2, C_3 follow in a similar way; so, finally, remembering that $\mathbf{C} = C_1\mathbf{i} + C_2\mathbf{j} + C_3\mathbf{k}$, we find that

$$\mathbf{C} = \nabla \times (\mathbf{A} \times \mathbf{B}) = (\nabla . \mathbf{B})\mathbf{A} - (\nabla . \mathbf{A})\mathbf{B} + (\mathbf{B} . \nabla)\mathbf{A} - (\mathbf{A} . \nabla)\mathbf{B}.$$

EXAMPLE 2. *Find an expression for* curl curl $\mathbf{V}$.

$$\text{curl curl}\,\mathbf{V} = \nabla \times (\nabla \times \mathbf{V})$$

$$= \begin{vmatrix} \mathbf{i} & \mathbf{j} & \mathbf{k} \\ \dfrac{\partial}{\partial x} & \dfrac{\partial}{\partial y} & \dfrac{\partial}{\partial z} \\ \dfrac{\partial V_3}{\partial y} - \dfrac{\partial V_2}{\partial z}, & \dfrac{\partial V_1}{\partial z} - \dfrac{\partial V_3}{\partial x}, & \dfrac{\partial V_2}{\partial x} - \dfrac{\partial V_1}{\partial y} \end{vmatrix},$$

or

$$\text{curl curl}\,\mathbf{V} = \mathbf{C} = C_1\mathbf{i} + C_2\mathbf{j} + C_3\mathbf{k}, \text{ say,}$$

where

$$C_1 = \frac{\partial}{\partial y}\left(\frac{\partial V_2}{\partial x} - \frac{\partial V_1}{\partial y}\right) - \frac{\partial}{\partial z}\left(\frac{\partial V_1}{\partial z} - \frac{\partial V_3}{\partial x}\right)$$

$$= \frac{\partial^2 V_2}{\partial y\,\partial x} + \frac{\partial^2 V_3}{\partial z\,\partial x} - \frac{\partial^2 V_1}{\partial y^2} - \frac{\partial^2 V_1}{\partial z^2}$$

$$= \left(\frac{\partial^2 V_1}{\partial x^2} + \frac{\partial^2 V_2}{\partial x\,\partial y} + \frac{\partial^2 V_3}{\partial x\,\partial z}\right) - \left(\frac{\partial^2 V_1}{\partial x^2} + \frac{\partial^2 V_1}{\partial y^2} + \frac{\partial^2 V_1}{\partial z^2}\right)$$

$$= \frac{\partial}{\partial x}\left(\frac{\partial V_1}{\partial x} + \frac{\partial V_2}{\partial y} + \frac{\partial V_3}{\partial z}\right) - \nabla^2 V_1.$$

So

$$C_1 = \frac{\partial}{\partial x}(\nabla \,.\, \mathbf{V}) - \nabla^2 V_1,$$

with two similar expressions for C_2 and C_3. Hence

$$\mathbf{C} = \left(\mathbf{i}\frac{\partial}{\partial x} + \mathbf{j}\frac{\partial}{\partial y} + \mathbf{k}\frac{\partial}{\partial z}\right)\left(\nabla \,.\, \mathbf{V}\right) - (\nabla^2 V_1\mathbf{i} + \nabla^2 V_2\mathbf{j} + \nabla^2 V_3\mathbf{k}).$$

Thus

$$\text{curl curl}\,\mathbf{V} = \nabla(\nabla \,.\, \mathbf{V}) - \nabla^2\mathbf{V}, \qquad (15)$$

$$= \text{grad div}\,\mathbf{V} - \nabla^2\mathbf{V}.$$

The reader should again note the result of operating with the scalar operator ∇^2 on a vector $\mathbf{V}$; ∇^2 acts on each component of $\mathbf{V}$, producing

$$(\nabla^2 V_1)\mathbf{i} + (\nabla^2 V_2)\mathbf{j} + (\nabla^2 V_3)\mathbf{k}.$$

This has been mentioned earlier in § 6:8.

EXAMPLE 3. *Show that*

$$(\mathbf{c} \times \nabla) \times \mathbf{A} = \operatorname{grad}(\mathbf{A} \,.\, \mathbf{c}) - (\operatorname{div} \mathbf{A})\mathbf{c},$$

where $\mathbf{c}$ *is a constant vector and* $\mathbf{A}$ *a vector function of position.*

Let $\mathbf{c} = c_1\mathbf{i} + c_2\mathbf{j} + c_3\mathbf{k}$, $\mathbf{A} = A_1\mathbf{i} + A_2\mathbf{j} + A_3\mathbf{k}$. Then c_1, c_2, c_3 are all constants, and A_1, A_2, A_3 are all functions of x, y and z. $\mathbf{c} \times \nabla$ is a vector operator; for

$$\mathbf{c} \times \nabla = (c_1\mathbf{i} + c_2\mathbf{j} + c_3\mathbf{k}) \times \left(\mathbf{i}\frac{\partial}{\partial x} + \mathbf{j}\frac{\partial}{\partial y} + \mathbf{k}\frac{\partial}{\partial z}\right)$$

$$= \left(c_2\frac{\partial}{\partial z} - c_3\frac{\partial}{\partial y}\right)\mathbf{i} + \left(c_3\frac{\partial}{\partial x} - c_1\frac{\partial}{\partial z}\right)\mathbf{j} + \left(c_1\frac{\partial}{\partial y} - c_2\frac{\partial}{\partial x}\right)\mathbf{k}.$$

Let
$$(\mathbf{c} \times \nabla) \times \mathbf{A} = \mathbf{B} = B_1\mathbf{i} + B_2\mathbf{j} + B_3\mathbf{k}. \tag{16}$$

Then
$$B_1 = \left(c_3\frac{\partial}{\partial x} - c_1\frac{\partial}{\partial z}\right)A_3 - \left(c_1\frac{\partial}{\partial y} - c_2\frac{\partial}{\partial x}\right)A_2$$

$$= c_3\frac{\partial A_3}{\partial x} + c_2\frac{\partial A_2}{\partial x} - c_1\left(\frac{\partial A_2}{\partial y} + \frac{\partial A_3}{\partial z}\right).$$

So, adding and subtracting $c_1\dfrac{\partial A_1}{\partial x}$,

$$B_1 = \left(c_3\frac{\partial A_3}{\partial x} + c_2\frac{\partial A_2}{\partial x} + c_1\frac{\partial A_1}{\partial x}\right) - c_1\left(\frac{\partial A_1}{\partial x} + \frac{\partial A_2}{\partial y} + \frac{\partial A_3}{\partial z}\right).$$

Since $\mathbf{c}$ is a constant vector,

$$B_1 = \frac{\partial}{\partial x}(A_1c_1 + A_2c_2 + A_3c_3) - c_1(\operatorname{div} \mathbf{A}).$$

So
$$B_1 = \frac{\partial}{\partial x}(\mathbf{A} \,.\, \mathbf{c}) - c_1(\operatorname{div} \mathbf{A}).$$

Substituting for B_1 in (16), together with similar expressions for B_2, B_3, we find

$$(\mathbf{c} \times \nabla) \times \mathbf{A} = \left[\frac{\partial}{\partial x}(\mathbf{A} \cdot \mathbf{c})\,\mathbf{i} + \frac{\partial}{\partial y}(\mathbf{A} \cdot \mathbf{c})\,\mathbf{j} + \frac{\partial}{\partial z}(\mathbf{A} \cdot \mathbf{c})\,\mathbf{k}\right] -$$

$$- \operatorname{div} \mathbf{A}(c_1\mathbf{i} + c_2\mathbf{j} + c_3\mathbf{k}),$$

$$= \operatorname{grad}(\mathbf{A} \cdot \mathbf{c}) - (\operatorname{div} \mathbf{A})\mathbf{c}.$$

EXAMPLE 4. *If* $r^2 = x^2 + y^2 + z^2$, *show that*

$$\operatorname{curl}[\mathbf{k} \times \operatorname{grad}(1/r)] + \operatorname{grad}[\mathbf{k} \cdot \operatorname{grad}(1/r)] = \mathbf{0}.$$

First we find grad $(1/r)$.

$$\operatorname{grad}\left(\frac{1}{r}\right) = \left(\mathbf{i}\frac{\partial}{\partial x} + \mathbf{j}\frac{\partial}{\partial y} + \mathbf{k}\frac{\partial}{\partial z}\right)\left(\frac{1}{r}\right)$$

$$= -\frac{1}{r^2}\left(\frac{\partial r}{\partial x}\mathbf{i} + \frac{\partial r}{\partial y}\mathbf{j} + \frac{\partial r}{\partial z}\mathbf{k}\right);$$

thus
$$\operatorname{grad}\left(\frac{1}{r}\right) = -\frac{1}{r^3}(x\mathbf{i} + y\mathbf{j} + z\mathbf{k}).$$

So
$$\mathbf{k} \cdot \operatorname{grad}\left(\frac{1}{r}\right) = -\frac{1}{r^3}\mathbf{k} \cdot (x\mathbf{i} + y\mathbf{j} + z\mathbf{k});$$

therefore
$$\mathbf{k} \cdot \operatorname{grad}\left(\frac{1}{r}\right) = -\frac{z}{r^3}. \tag{17}$$

Also
$$\mathbf{k} \times \operatorname{grad}\left(\frac{1}{r}\right) = -\frac{1}{r^3}\mathbf{k} \times (x\mathbf{i} + y\mathbf{j} + z\mathbf{k})$$

$$= \frac{y}{r^3}\mathbf{i} - \frac{x}{r^3}\mathbf{j}. \tag{18}$$

Using (17),

$$\operatorname{grad}\left[\mathbf{k} \cdot \operatorname{grad}\left(\frac{1}{r}\right)\right] = \left(\mathbf{i}\frac{\partial}{\partial x} + \mathbf{j}\frac{\partial}{\partial y} + \mathbf{k}\frac{\partial}{\partial z}\right)\left(-\frac{z}{r^3}\right).$$

Thus
$$\operatorname{grad}\left[\mathbf{k} \cdot \operatorname{grad}\left(\frac{1}{r}\right)\right] = \frac{3xz}{r^5}\mathbf{i} + \frac{3yz}{r^5}\mathbf{j} + \left(\frac{3z^2}{r^5} - \frac{1}{r^3}\right)\mathbf{k}. \tag{19}$$

Using (18),

$$\operatorname{curl}\left[\mathbf{k} \times \operatorname{grad}\left(\frac{1}{r}\right)\right] = \begin{vmatrix} \mathbf{i} & \mathbf{j} & \mathbf{k} \\ \dfrac{\partial}{\partial x} & \dfrac{\partial}{\partial y} & \dfrac{\partial}{\partial z} \\ \dfrac{y}{r^3} & -\dfrac{x}{r^3} & 0 \end{vmatrix}$$

$$= \frac{\partial}{\partial z}\left(\frac{x}{r^3}\right)\mathbf{i} + \frac{\partial}{\partial z}\left(\frac{y}{r^3}\right)\mathbf{j} - \left[\frac{\partial}{\partial x}\left(\frac{x}{r^3}\right) + \frac{\partial}{\partial y}\left(\frac{y}{r^3}\right)\right]\mathbf{k}.$$

Therefore

$$\operatorname{curl}\left[\mathbf{k} \times \operatorname{grad}\left(\frac{1}{r}\right)\right] = -\frac{3xz}{r^5}\mathbf{i} - \frac{3yz}{r^5}\mathbf{j} - \left[\frac{2}{r^3} - \frac{3x^2}{r^5} - \frac{3y^2}{r^5}\right]\mathbf{k}$$

$$= -\frac{3xz}{r^5}\mathbf{i} - \frac{3yz}{r^5}\mathbf{j} - \left[-\frac{1}{r^3} + \frac{3z^2}{r^5}\right]\mathbf{k}. \qquad (20)$$

From (19) and (20) we obtain

$$\operatorname{curl}[\mathbf{k} \times \operatorname{grad}(1/r)] + \operatorname{grad}[\mathbf{k} \cdot \operatorname{grad}(1/r)] = \mathbf{0}.$$

Note that we could have used the result (14) of Example 1 to expand

$$\operatorname{curl}[\mathbf{k} \times \operatorname{grad}(1/r)].$$

Exercises 7:5

1. Scalar point functions ϕ, ψ are defined by the equations $\phi = \mathbf{a} \cdot \nabla r^3$, $\psi = \mathbf{b} \cdot \nabla(1/r^3)$, where **a** and **b** are constant vectors. Evaluate
   (i) $\mathbf{b} \cdot \operatorname{curl}(\phi\mathbf{r}) - r^6\mathbf{a} \cdot \operatorname{curl}(\psi\mathbf{r})$,
   (ii) $\mathbf{b} \cdot \operatorname{curl}[\operatorname{div}(r^2\mathbf{r})(\mathbf{a} \times \mathbf{r})] + 5r^4\mathbf{a} \cdot \operatorname{curl}[\operatorname{div}(r^{-2}\mathbf{r})(\mathbf{b} \times \mathbf{r})]$,
   (iii) $\operatorname{curl}(\phi\psi\mathbf{r})$.
2. If **A** is a vector point function and **c** is a constant vector, show that

   $$\mathbf{c} \cdot \operatorname{curl}\operatorname{curl}\mathbf{A} = \mathbf{c} \cdot \operatorname{grad}\operatorname{div}\mathbf{A} - \operatorname{div}\operatorname{grad}(\mathbf{c} \cdot \mathbf{A}). \qquad \text{[B.Sc.(Gen.)]}$$

3. Establish the identity

   $$\operatorname{grad}(\mathbf{a} \cdot \mathbf{b}) = \mathbf{a} \times \operatorname{curl}\mathbf{b} + \mathbf{b} \times \operatorname{curl}\mathbf{a} + (\mathbf{a} \cdot \nabla)\mathbf{b} + (\mathbf{b} \cdot \nabla)\mathbf{a}$$

for any two vector point functions **a** and **b**. Hence show that

$$(\mathbf{u}\,.\,\nabla)\mathbf{u} = \tfrac{1}{2}\,\text{grad}\,(\mathbf{u}^2) - \mathbf{u} \times \text{curl}\,\mathbf{u},$$

for any vector point function **u**.

If $\mathbf{u} = -\,\text{curl}\,\phi\,\mathbf{k}$, where ϕ is a function of x and y, prove that $\text{curl}\,(\mathbf{u}\,.\,\nabla)\mathbf{u} = (\mathbf{u}\,.\,\nabla)\,\text{curl}\,\mathbf{u} = \text{grad}\,\phi \times \text{grad}\,\nabla^2\phi$. [B.Sc.(Eng.)]

7:6 Orthogonal curvilinear coordinates

So far we have been working in the rectangular cartesian coordinate system; now we consider a more general system of coordinates.

We begin with the familiar cartesian coordinates x, y, z and transform them by means of equations of the type

$$x = x(u,v,w), \quad y = y(u,v,w), \quad z = z(u,v,w), \tag{1}$$

to a new coordinate system u, v, w. We shall assume that the equations (1) can be solved to give u, v, w as functions of x, y, z. Then

$$u = u(x,y,z), \quad v = v(x,y,z), \quad w = w(x,y,z). \tag{2}$$

We assume also that equations (1) and (2) are such that to any point (x, y, z) corresponds just one set of values of u, v, w. It may, of course, be necessary to restrict the ranges of values for u, v, w to achieve this; see, for example, the spherical polar system discussed in § 7:8 (i).

The equation $u(x, y, z) =$ constant represents a surface in three-dimensional space; as we change the constant we get a family of level surfaces. Since to any point (x, y, z) there corresponds *one* value only of u, it follows that *one surface only* of the family $u =$ constant, passes through any given point $P(x, y, z)$. Similarly, *one* surface of the family $v =$ constant and *one* surface of the family $w =$ constant pass through P. Thus *three* surfaces, one of each family, pass through any point $P(x, y, z)$; these are the *coordinate surfaces*.

In Fig. 38, the area $PRQ'S$ is part of the surface $u =$ constant, passing through P. Similarly, $PQR'S$, $PQS'R$ are parts of the surfaces $v =$ constant, $w =$ constant. (The surfaces extend beyond

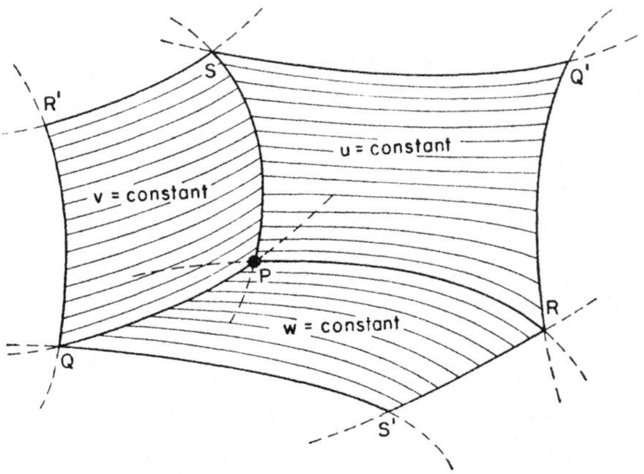

FIG. 38

P in each case; this is indicated by the dotted lines in the diagram.) It must be stressed that the coordinate surfaces are, in general, *curved* surfaces, although in particular cases one or more may be planes; in the spherical polar system, for example, the three families of surfaces are spheres, cones and planes. The rectangular cartesian system is the very special case obtained by writing $x = u$, $y = v$, $z = w$, and all three families of coordinate surfaces are planes.

The coordinate surfaces at the point $P(x, y, z)$ meet in three lines, *the coordinate lines*, through P; these are the lines PQ, PR, PS in Fig. 38. In general these are *not* straight lines. If $P(x, y, z)$ is a given point in the cartesian system, then the values of u, v, w corresponding to P are the *curvilinear coordinates* of P.

We shall consider here only the important case in which the coordinate surfaces at P are mutually orthogonal. It follows that the coordinate lines are mutually orthogonal at P. Suppose that PQ, PR, PS are the coordinate lines at $P(u, v, w)$. Assume that Q, R, S are near to P, so that Q is the point $(u + \mathrm{d}u, v, w)$, R the point $(u, v + \mathrm{d}v, w)$, and S the point $(u, v, w + \mathrm{d}w)$. Then PQ is the curve of intersection of the surfaces $v =$ constant, $w =$ constant, so that on PQ only u varies. Similarly, on PR only v varies and on PS only w varies. The lines PQ, PR, PS are shown in Fig. 39. We shall take $\hat{\mathbf{a}}$, $\hat{\mathbf{b}}$, $\hat{\mathbf{c}}$ as unit vectors in the directions of

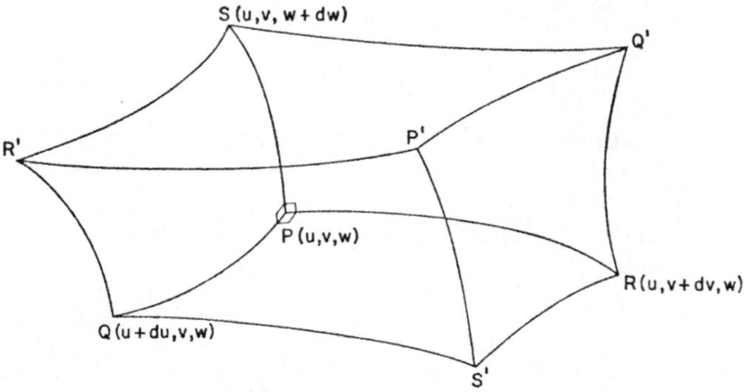

FIG. 39

PQ, PR, PS at P, and we shall assume throughout that they form a right-handed set. The arc-length PQ is the element of arc along the coordinate line PQ, and we denote it by $\mathrm{d}s_1$. Similarly, we write $PR = \mathrm{d}s_2$, $PS = \mathrm{d}s_3$. We can complete the element of volume at P by inserting the points $P'(u + \mathrm{d}u, v + \mathrm{d}v, w + \mathrm{d}w)$, $Q'(u, v + \mathrm{d}v, w + \mathrm{d}w)$, $R'(u + \mathrm{d}u, v, w + \mathrm{d}w)$ and $S'(u + \mathrm{d}u, v + \mathrm{d}v, w)$. The reader must note particularly that in a general orthogonal curvilinear system we cannot assume equality between lengths such as PR and QS', or between PS and QR'; they may be equal in special cases (see § 7:8) but they are not equal in general.

Our first task is to find expressions for $\mathrm{d}s_1$, $\mathrm{d}s_2$, $\mathrm{d}s_3$, in terms of u, v, w.

Since by (1) $x = x(u, v, w)$,

$$\mathrm{d}x = \frac{\partial x}{\partial u}\mathrm{d}u + \frac{\partial x}{\partial v}\mathrm{d}v + \frac{\partial x}{\partial w}\mathrm{d}w. \tag{3a}$$

Similarly,

$$\mathrm{d}y = \frac{\partial y}{\partial u}\mathrm{d}u + \frac{\partial y}{\partial v}\mathrm{d}v + \frac{\partial y}{\partial w}\mathrm{d}w, \tag{3b}$$

$$dz = \frac{\partial z}{\partial u}du + \frac{\partial z}{\partial v}dv + \frac{\partial z}{\partial w}dw. \tag{3c}$$

Let the element of length be denoted by dl; then

$$(dl)^2 = (dx)^2 + (dy)^2 + (dz)^2,$$

and we can substitute for $(dx)^2$, $(dy)^2$, $(dz)^2$ from equations (3). Writing dx^2 for $(dx)^2$, etc., we find

$$dl^2 = g_{11}\,du^2 + g_{22}\,dv^2 + g_{33}\,dw^2 + 2g_{23}\,dv\,dw + \\ + 2g_{31}\,dw\,du + 2g_{12}\,du\,dv, \tag{4}$$

where
$$\left.\begin{aligned} g_{11} &= \left(\frac{\partial x}{\partial u}\right)^2 + \left(\frac{\partial y}{\partial u}\right)^2 + \left(\frac{\partial z}{\partial u}\right)^2, \text{ etc.,} \\ g_{12} &= \frac{\partial x}{\partial u}\frac{\partial x}{\partial v} + \frac{\partial y}{\partial u}\frac{\partial y}{\partial v} + \frac{\partial z}{\partial u}\frac{\partial z}{\partial v}, \text{ etc.} \end{aligned}\right\} \tag{5}$$

We are dealing only with orthogonal curvilinear systems, and for these $g_{12} = g_{23} = g_{31} = 0$ always. [See M. R. Spiegel, *Vector Analysis*, Schaum, New York, 1959, chapter 7.] In (4) let us now keep v and w constant. Then dv and dw are both zero, and the line-element dl reduces to the element of length ds_1 along the coordinate line PQ. So

$$ds_1 = \sqrt{g_{11}}\,.\,du.$$

Similarly, $\quad ds_2 = \sqrt{g_{22}}\,.\,dv, \quad ds_3 = \sqrt{g_{33}}\,.\,dw.$

It is usual to write these results as

$$ds_1 = h_1\,du, \quad ds_2 = h_2\,dv, \quad ds_3 = h_3\,dw. \tag{6}$$

We next find expressions for the elements of area. In the cartesian system the element of area in the plane x = constant is $dy\,dz$; corresponding to this in the curvilinear system the element of area on the surface u = constant is the area $PRQ'S$ in Fig. 39, which to a first approximation is given by $PR \times PS$.

So when du, dv, dw are infinitesimally small the element of area on u = constant may be taken as

$$ds_2\,ds_3 = h_2 h_3\,dv\,dw. \tag{7}$$

In a similar way, the elements of area on the surfaces v = constant, w = constant are $h_3 h_1\,dw\,du$, $h_1 h_2\,du\,dv$ respectively.

Finally the element of volume in this system is the volume $PQS'RQ'SR'P'$. To a first approximation this is given by

$$ds_1\,ds_2\,ds_3 = h_1 h_2 h_3\,du\,dv\,dw. \tag{8}$$

Thus (8) may be taken as giving the element of volume when du, dv, dw are infinitesimally small.

In particular coordinate systems, such as that of spherical polars for example, the quantities h_1, h_2, h_3 can easily be found. See § 7:8 (i), where h_1, h_2, h_3 are evaluated.

Exercises 7:6

1. For each of the following orthogonal curvilinear systems evaluate the coefficients g_{ij} ($i = 1, 2, 3$, $j = 1, 2, 3$), and hence write down the expressions for dl^2 and for the element of volume:
   (i) the spherical polar system $x = r\sin\theta\cos\psi$, $y = r\sin\theta\sin\psi$, $z = r\cos\theta$;
   (ii) the cylindrical polar system $x = \rho\cos\psi$, $y = \rho\sin\psi$, $z = z$;
   (iii) the paraboloidal system $x = uv\cos\psi$, $y = uv\sin\psi$, $z = \frac{1}{2}(u^2 - v^2)$.

7:7 Gradient, divergence and curl in curvilinear systems

(i) grad ϕ *in curvilinear coordinates*

The reader will recall that grad ϕ is defined in terms of the rectangular coordinates by

$$\text{grad}\,\phi = \frac{\partial\phi}{\partial x}\mathbf{i} + \frac{\partial\phi}{\partial y}\mathbf{j} + \frac{\partial\phi}{\partial z}\mathbf{k}.$$

In the curvilinear system, $\hat{\mathbf{a}}$, $\hat{\mathbf{b}}$, $\hat{\mathbf{c}}$ take the place of $\mathbf{i}$, $\mathbf{j}$, $\mathbf{k}$, and $\partial/\partial x$ is replaced by $\partial/\partial s_1$, $\partial/\partial y$ by $\partial/\partial s_2$, $\partial/\partial z$ by $\partial/\partial s_3$; this is so since ds_1 is the element of length in the direction of the unit vector $\hat{\mathbf{a}}$, just as dx is the element of length in the direction of the unit vector $\mathbf{i}$, and similarly for ds_2, ds_3. Hence

$$\operatorname{grad}\phi = \frac{\partial\phi}{\partial s_1}\hat{\mathbf{a}} + \frac{\partial\phi}{\partial s_2}\hat{\mathbf{b}} + \frac{\partial\phi}{\partial s_3}\hat{\mathbf{c}}.$$

Using 7:6(6), this becomes

$$\operatorname{grad}\phi = \frac{1}{h_1}\frac{\partial\phi}{\partial u}\hat{\mathbf{a}} + \frac{1}{h_2}\frac{\partial\phi}{\partial v}\hat{\mathbf{b}} + \frac{1}{h_3}\frac{\partial\phi}{\partial w}\hat{\mathbf{c}}. \tag{1}$$

Thus the components of grad ϕ in the direction of the coordinate lines at $P(u, v, w)$ are

$$\frac{1}{h_1}\frac{\partial\phi}{\partial u},\quad \frac{1}{h_2}\frac{\partial\phi}{\partial v},\quad \frac{1}{h_3}\frac{\partial\phi}{\partial w}. \tag{2}$$

The reader should note that whereas in the cartesian system the vectors $\mathbf{i}$, $\mathbf{j}$, $\mathbf{k}$ are constant in direction for every point of the scalar field ϕ, in the curvilinear system $\hat{\mathbf{a}}$, $\hat{\mathbf{b}}$, $\hat{\mathbf{c}}$ *change* in direction from point to point. (We shall illustrate this in § 7:8 for the special cases of spherical and cylindrical polar coordinates.) In fact we can write

$$\hat{\mathbf{a}} = a_1\mathbf{i} + a_2\mathbf{j} + a_3\mathbf{k},$$

where a_1, a_2, a_3 are each functions of u, v, w, such that

$$a_1^2 + a_2^2 + a_3^2 = 1, \text{ since } \hat{\mathbf{a}} \text{ is a unit vector.}$$

(ii) div $\mathbf{V}$ *in curvilinear coordinates*

Suppose that at the point $P(u, v, w)$ the vector $\mathbf{V}$ has components $V_1(u, v, w)$, $V_2(u, v, w)$, $V_3(u, v, w)$ in the directions of the coordinate lines at P. Then we shall show that

$$\operatorname{div}\mathbf{V} = \frac{1}{h_1h_2h_3}\left[\frac{\partial}{\partial u}(h_2h_3V_1) + \frac{\partial}{\partial v}(h_3h_1V_2) + \frac{\partial}{\partial w}(h_1h_2V_3)\right]. \tag{3}$$

We shall use the integral definition of the divergence to establish this result. Readers interested only in applications may assume this result and go on to the sections on particular coordinate systems. [We can of course write

$$\nabla = \frac{1}{h_1}\hat{\mathbf{a}}\frac{\partial}{\partial u} + \frac{1}{h_2}\hat{\mathbf{b}}\frac{\partial}{\partial v} + \frac{1}{h_3}\hat{\mathbf{c}}\frac{\partial}{\partial w}$$

for the vector operator ∇ in the curvilinear system. Then

$$\operatorname{div} \mathbf{V} = \left(\frac{1}{h_1}\hat{\mathbf{a}}\frac{\partial}{\partial u} + \frac{1}{h_2}\hat{\mathbf{b}}\frac{\partial}{\partial v} + \frac{1}{h_3}\hat{\mathbf{c}}\frac{\partial}{\partial w}\right) \cdot (V_1\hat{\mathbf{a}} + V_2\hat{\mathbf{b}} + V_3\hat{\mathbf{c}}).$$

The value of div **V** does not follow immediately from this since the differential operators $\partial/\partial u$, $\partial/\partial v$, $\partial/\partial w$ now operate *not only* on V_1, V_2, V_3, *but also* on the unit vectors $\hat{\mathbf{a}}$, $\hat{\mathbf{b}}$, $\hat{\mathbf{c}}$; for these unit vectors vary in direction from point to point and their components are functions of u, v and w. For this method, see M. R. Spiegel, *Vector Analysis*, Schaum, New York, 1959.]

To use the integral definition of the divergence, we have to find the flux of **V** through a small volume and then take the limiting value of the ratio of this flux to the volume as the volume tends to zero. We shall evaluate the divergence at the point $P(u, v, w)$,

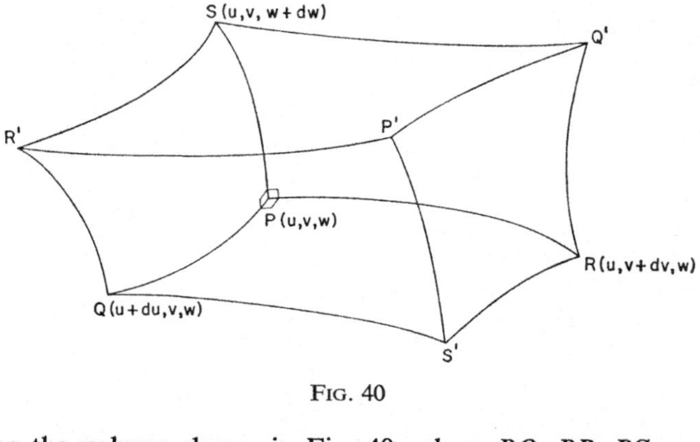

FIG. 40

using the volume shown in Fig. 40, where PQ, PR, PS are the elements of length along the coordinate lines at P. So $PQ = ds_1$,

$PR = ds_2$, $PS = ds_3$. Since we are dealing with the general orthogonal curvilinear system, opposite edges of the volume element cannot be assumed of equal length. This means that though we take the area of $PRQ'S$ as $ds_2 ds_3$ to a first approximation, we *cannot* assume this for the area $QS'P'R'$ to the same degree of approximation. The simplest way to deal with this problem is to consider, as in § 6:7, the resultant flux through pairs of faces of the volume. The flux through the face $PQS'R$ is by definition

$$\mathbf{V}(P) \,.\, d\mathbf{S} = (V_1\hat{\mathbf{a}} + V_2\hat{\mathbf{b}} + V_3\hat{\mathbf{c}}) \,.\, (-\, ds_1\, ds_2\, \hat{\mathbf{c}})$$

$$= -\, V_3(u, v, w)\, ds_1\, ds_2 \,.$$

Therefore

$$\mathbf{V}(P) \,.\, d\mathbf{S} = -\, V_3 h_1 h_2\, du\, dv. \tag{4}$$

For the face $SR'P'Q'$ we shall have a similar expression, with a positive sign, and $V_3h_1h_2$ replaced by its corresponding value on this face; but here u, v are unchanged and w has changed its value to $w + dw$. Writing $V_3h_1h_2$ as $f(u, v, w)$, what we now require is $f(u, v, w + dw)$. This we can obtain by Taylor's Theorem. So

$$f(u, v, w + dw) = f(u, v, w) + \frac{\partial}{\partial w}\{ f(u, v, w)\}\, dw$$

$$= V_3 h_1 h_2 + \frac{\partial}{\partial w}(V_3 h_1 h_2)\, dw,$$

ignoring terms of higher order in dw. Hence the flux through the face $SR'P'Q'$ is

$$\left[V_3 h_1 h_2 + \frac{\partial}{\partial w}(V_3 h_1 h_2)\, dw \right] du\, dv. \tag{5}$$

Combining (4) and (5), the contribution to the total flux from the pair of faces $PQS'R$, $SR'P'Q'$ is

$$\delta F_3 = \frac{\partial}{\partial w}(V_3 h_1 h_2)\,du\,dv\,dw. \tag{6a}$$

Similarly, we obtain the two other contributions

$$\delta F_1 = \frac{\partial}{\partial u}(V_1 h_2 h_3)\,du\,dv\,dw, \tag{6b}$$

$$\delta F_2 = \frac{\partial}{\partial v}(V_2 h_3 h_1)\,du\,dv\,dw. \tag{6c}$$

The total flux through the volume element is found by adding together δF_1, δF_2, δF_3, as given by equations (6). We now divide this total flux by the element of volume $ds_1\,ds_2\,ds_3$, that is by $h_1 h_2 h_3\,du\,dv\,dw$, and then take the limiting value as the volume shrinks to zero. Thus we get

$$\operatorname{div}\mathbf{V} = \frac{1}{h_1\,h_2\,h_3}\left[\frac{\partial}{\partial u}(V_1 h_2 h_3) + \frac{\partial}{\partial v}(V_2 h_3 h_1) + \frac{\partial}{\partial w}(V_3 h_1 h_2)\right]. \tag{7}$$

It is now easy to evaluate $\nabla^2\phi$ in curvilinear coordinates.

Since $$\nabla^2\phi = \operatorname{div}\operatorname{grad}\phi$$

we take $\mathbf{V} = \operatorname{grad}\phi$, so that by (2)

$$V_1 = \frac{1}{h_1}\frac{\partial\phi}{\partial u}, \quad V_2 = \frac{1}{h_2}\frac{\partial\phi}{\partial v}, \quad V_3 = \frac{1}{h_3}\frac{\partial\phi}{\partial w}.$$

Then, applying (7),

$$\nabla^2\phi = \frac{1}{h_1 h_2 h_3}\left[\frac{\partial}{\partial u}\left(\frac{h_2 h_3}{h_1}\frac{\partial\phi}{\partial u}\right) + \frac{\partial}{\partial v}\left(\frac{h_3 h_1}{h_2}\frac{\partial\phi}{\partial v}\right) + \frac{\partial}{\partial w}\left(\frac{h_1 h_2}{h_3}\frac{\partial\phi}{\partial w}\right)\right]. \tag{8}$$

We shall return to this expression in § 7:9.

(iii) curl **V** *in curvilinear coordinates*

We shall show that the components of curl **V** in the direction of the coordinate lines at $P(u, v, w)$ are

$$\frac{1}{h_2h_3}\left[\frac{\partial}{\partial v}(h_3V_3)-\frac{\partial}{\partial w}(h_2V_2)\right],\quad \frac{1}{h_3h_1}\left[\frac{\partial}{\partial w}(h_1V_1)-\frac{\partial}{\partial u}(h_3V_3)\right],$$

$$\frac{1}{h_1h_2}\left[\frac{\partial}{\partial u}(h_2V_2)-\frac{\partial}{\partial v}(h_1V_1)\right]. \tag{9}$$

Again, readers interested only in applications may omit the following analysis.

To establish (9) we use the integral definition of curl **V**. This means that to find the component in the direction of the unit vector $\hat{\mathbf{a}}$ at P, we have first to find the circulation in the small circuit $PRQ'S$, say (see Fig. 41), then divide by the area of $PRQ'S$

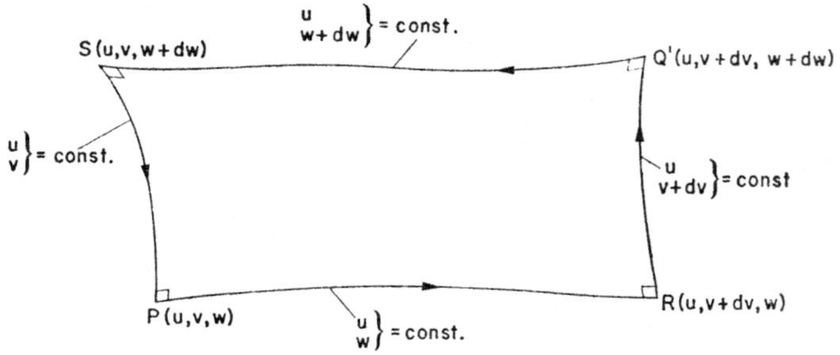

FIG. 41

and take the limit as this area tends to zero. Denoting the circulation by δC, we have by definition

$$\delta C = \int_{PRQ'S} \mathbf{V} \,.\, d\mathbf{r}$$

$$= \int_{PR} \mathbf{V} \,.\, d\mathbf{r} + \int_{RQ'} \mathbf{V} \,.\, d\mathbf{r} + \int_{Q'S} \mathbf{V} \,.\, d\mathbf{r} + \int_{SP} \mathbf{V} \,.\, d\mathbf{r}. \tag{10}$$

Consider $\int_{PR} \mathbf{V} \cdot \mathrm{d}\mathbf{r}$; on PR only v changes, so that $\mathrm{d}\mathbf{r} = \mathrm{d}s_2\hat{\mathbf{b}}$. Thus

$$\begin{aligned}\mathbf{V} \cdot \mathrm{d} &= (V_1\hat{\mathbf{a}} + V_2\hat{\mathbf{b}} + V_3\hat{\mathbf{c}}) \cdot \mathrm{d}s_2\hat{\mathbf{b}} \\ &= V_2\,\mathrm{d}s_2.\end{aligned}$$

Since PR is of length $\mathrm{d}s_2$, we now get

$$\int_{PR} \mathbf{V} \cdot \mathrm{d}\mathbf{r} = V_2\,\mathrm{d}s_2 = V_2h_2\,\mathrm{d}v. \tag{11}$$

Now consider $\int_{Q'S} \mathbf{V} \cdot \mathrm{d}\mathbf{r}$; this is $-\int_{SQ'} \mathbf{V} \cdot \mathrm{d}\mathbf{r}$, and will have the same form as (11) with V_2h_2 replaced by its proper value on SQ'. But SQ' is given by the intersection of the surfaces u = constant, $(w + \mathrm{d}w)$ = constant. Thus, if we write $V_2h_2 = f(u, v, w)$, what we require is $f(u, v, w + \mathrm{d}w)$. To the first order in $\mathrm{d}w$ this is

$$f(u, v, w) + \frac{\partial}{\partial w} f(u, v, w)\,\mathrm{d}w.$$

Thus

$$-\int_{SQ'} \mathbf{V} \cdot \mathrm{d}\mathbf{r} = -\left[V_2h_2 + \frac{\partial}{\partial w}(V_2h_2)\,\mathrm{d}w\right]\mathrm{d}v. \tag{12}$$

Take next $\int_{SP} \mathbf{V} \cdot \mathrm{d}\mathbf{r}$; on PS only w changes, so that here $\mathrm{d}\mathbf{r} = \mathrm{d}s_3\hat{\mathbf{c}}$. So

$$\begin{aligned}\mathbf{V} \cdot \mathrm{d}\mathbf{r} &= (V_1\hat{\mathbf{a}} + V_2\hat{\mathbf{b}} + V_3\hat{\mathbf{c}}) \cdot \mathrm{d}s_3\hat{\mathbf{c}} \\ &= V_3\,\mathrm{d}s_3.\end{aligned}$$

Hence

$$\int_{SP} \mathbf{V} \cdot \mathrm{d}\mathbf{r} = -\int_{PS} V_3\,\mathrm{d}s_3 = -V_3\,\mathrm{d}s_3 = -V_3h_3\,\mathrm{d}w. \tag{13}$$

Finally, considering $\int_{RQ'} \mathbf{V} \cdot \mathrm{d}\mathbf{r}$ in the same way as $\int_{SQ'} \mathbf{V} \cdot \mathrm{d}\mathbf{r}$ we find

$$\int_{RQ'} \mathbf{V} \cdot \mathrm{d}\mathbf{r} = \left[V_3h_3 + \frac{\partial}{\partial v}(V_3h_3)\,\mathrm{d}v\right]\mathrm{d}w, \tag{14}$$

since RQ' is the intersection of the surfaces u = constant,

$(v + dv) =$ constant. Using (11), (12), (13) and (14) we find

$$\delta C = \int_{PRQ'S} \mathbf{V} \cdot d\mathbf{r}$$

$$= V_2 h_2 \, dv - \left[V_2 h_2 + \frac{\partial}{\partial w}(V_2 h_2) \, dw \right] dv +$$

$$+ \left[V_3 h_3 + \frac{\partial}{\partial v}(V_3 h_3) \, dv \right] dw - V_3 h_3 \, dw;$$

thus $$\delta C = \frac{\partial}{\partial v}(V_3 h_3) \, dv \, dw - \frac{\partial}{\partial w}(V_2 h_2) \, dv \, dw.$$

We now divide by $ds_2 ds_3$, that is by $h_2 h_3 dv dw$, and take the limit as $dv \, dw \to 0$. Thus

$$(\text{curl}\, \mathbf{V})_a = \frac{1}{h_2 h_3} \left[\frac{\partial}{\partial v}(V_3 h_3) - \frac{\partial}{\partial w}(V_2 h_2) \right].$$

The other two components follow immediately from symmetry.

We note here two identities already proved in § 7:3 for the cartesian coordinate system which can be shown to hold for any orthogonal curvilinear system. These are

$$\text{curl grad}\, \phi \equiv \mathbf{0}, \quad \text{div curl}\, \mathbf{V} \equiv 0.$$

We shall prove the first of these and leave the second for the reader as a simple exercise in the use of the formulae for div and curl given by (3) and (9). In the notation of the general curvilinear system we have

$$\text{grad}\, \phi = \frac{1}{h_1} \frac{\partial \phi}{\partial u} \hat{\mathbf{a}} + \frac{1}{h_2} \frac{\partial \phi}{\partial v} \hat{\mathbf{b}} + \frac{1}{h_3} \frac{\partial \phi}{\partial w} \hat{\mathbf{c}}.$$

So we require to find curl $\mathbf{V}$, where the components of $\mathbf{V}$ are

$$V_1 = \frac{1}{h_1} \frac{\partial \phi}{\partial u}, \quad V_2 = \frac{1}{h_2} \frac{\partial \phi}{\partial v}, \quad V_3 = \frac{1}{h_3} \frac{\partial \phi}{\partial w}.$$

Applying the formula given by (9), it follows that the component of curl **V** in the direction of $\hat{\mathbf{a}}$ is

$$\frac{1}{h_2 h_3}\left[\frac{\partial}{\partial v}\left(\frac{\partial \phi}{\partial w}\right) - \frac{\partial}{\partial w}\left(\frac{\partial \phi}{\partial v}\right)\right],$$

which is identically zero. Similarly, the other two components vanish, so that curl **V** is identically zero.

7:8 Two important curvilinear systems

We shall consider now two of the most frequently-used orthogonal curvilinear coordinate systems.

(i) *Spherical polar coordinates*

We transform from the rectangular cartesian coordinates x, y, z to spherical polar coordinates by means of the equations

$$x = r\sin\theta\cos\psi, \quad y = r\sin\theta\sin\psi, \quad z = r\cos\theta, \tag{1}$$

where $r \geqq 0, \quad 0 \leqq \theta < \pi, \quad 0 \leqq \psi < 2\pi$.

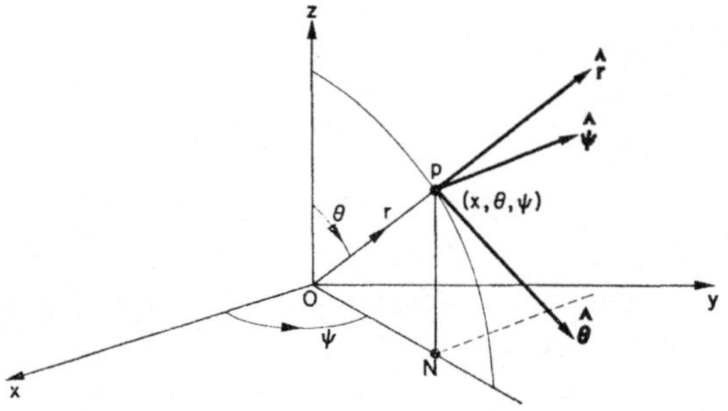

FIG. 42

Thus in Fig. 42 the distance OP is equal to r, while θ is measured from the z-axis towards OP and ψ lies in the x–y plane and is measured from the x-axis towards ON, where N is the foot of the perpendicular from P to the x–y plane.

The coefficients g_{ij} of 7:6(5) are easily calculated; for, replacing u, v, w of the theory by r, θ, ψ respectively, we have

$$\frac{\partial x}{\partial r} = \sin\theta\cos\psi, \quad \frac{\partial y}{\partial r} = \sin\theta\sin\psi, \quad \frac{\partial z}{\partial r} = \cos\theta,$$

$$\frac{\partial x}{\partial \theta} = r\cos\theta\cos\psi, \quad \frac{\partial y}{\partial \theta} = r\cos\theta\sin\psi, \quad \frac{\partial z}{\partial \theta} = -r\sin\theta,$$

$$\frac{\partial x}{\partial \psi} = -r\sin\theta\sin\psi, \quad \frac{\partial y}{\partial \psi} = r\sin\theta\cos\psi, \quad \frac{\partial z}{\partial \psi} = 0.$$

Hence
$$g_{11} = \left(\frac{\partial x}{\partial r}\right)^2 + \left(\frac{\partial y}{\partial r}\right)^2 + \left(\frac{\partial z}{\partial r}\right)^2 = 1,$$

$$g_{22} = \left(\frac{\partial x}{\partial \theta}\right)^2 + \left(\frac{\partial y}{\partial \theta}\right)^2 + \left(\frac{\partial z}{\partial \theta}\right)^2 = r^2,$$

$$g_{33} = \left(\frac{\partial x}{\partial \psi}\right)^2 + \left(\frac{\partial y}{\partial \psi}\right)^2 + \left(\frac{\partial z}{\partial \psi}\right)^2 = r^2\sin^2\theta.$$

Also
$$g_{12} = \frac{\partial x}{\partial r}\frac{\partial x}{\partial \theta} + \frac{\partial y}{\partial r}\frac{\partial y}{\partial \theta} + \frac{\partial z}{\partial r}\frac{\partial z}{\partial \theta} = 0,$$

$$g_{23} = \frac{\partial x}{\partial \theta}\frac{\partial x}{\partial \psi} + \frac{\partial y}{\partial \theta}\frac{\partial y}{\partial \psi} + \frac{\partial z}{\partial \theta}\frac{\partial z}{\partial \psi} = 0,$$

$$g_{31} = \frac{\partial x}{\partial \psi}\frac{\partial x}{\partial r} + \frac{\partial y}{\partial \psi}\frac{\partial y}{\partial r} + \frac{\partial z}{\partial \psi}\frac{\partial z}{\partial r} = 0,$$

in agreement with the statement in § 7:6. Substituting in equation (4) of § 7:6, we find

$$\mathrm{d}l^2 = \mathrm{d}r^2 + r^2\,\mathrm{d}\theta^2 + r^2\sin^2\theta\,\mathrm{d}\psi^2, \tag{2}$$

and also from 7:6(6)

$$h_1 = 1, \quad h_2 = r, \quad h_3 = r\sin\theta. \tag{3}$$

The element of volume in this system is thus

$$h_1 h_2 h_3\,\mathrm{d}u\,\mathrm{d}v\,\mathrm{d}w = r^2\sin\theta\,\mathrm{d}r\,\mathrm{d}\theta\,\mathrm{d}\psi. \tag{4}$$

The reader should recognise this form; it occurred in 4:7(2) in the change of variables to spherical polars in the evaluation of triple integrals.

The coordinate surfaces are easy to obtain. From (1), $r^2 = x^2 + y^2 + z^2$; thus r = constant gives a family of concentric spheres, centre the origin. Also $\tan^2\theta = (x^2 + y^2)/z^2$; so θ = constant gives the family of surfaces $x^2 + y^2 = c^2z^2$, where c is a constant. These surfaces are cones, vertex the origin. Finally, $\tan\psi = y/x$; so ψ = constant gives $y = cx$, which represents a family of planes through the z-axis. (In this case it is easy to see what the coordinate surfaces must be from a diagram, but in many systems it is necessary to find them analytically, as we have just done above; see Nos. 3 and 4 in Exercises 7:8 and No. 14 in Miscellaneous Exercises 7.)

At the point $P(r, \theta, \psi)$, we take unit vectors $\hat{\mathbf{r}}$, $\hat{\boldsymbol{\theta}}$, $\hat{\boldsymbol{\psi}}$ in the directions of r, θ, ψ increasing, as in Fig. 42. These unit vectors replace the $\hat{\mathbf{a}}$, $\hat{\mathbf{b}}$, $\hat{\mathbf{c}}$ of the general theory. Using our earlier notation for the volume element, we find from 7:6(6) that

$$PQ = \mathrm{d}s_1 = \mathrm{d}r, \quad PR = \mathrm{d}s_2 = r\,\mathrm{d}\theta, \quad PS = \mathrm{d}s_3 = r\sin\theta\,\mathrm{d}\psi.$$

The volume element is shown in Fig. 43. The reader is advised

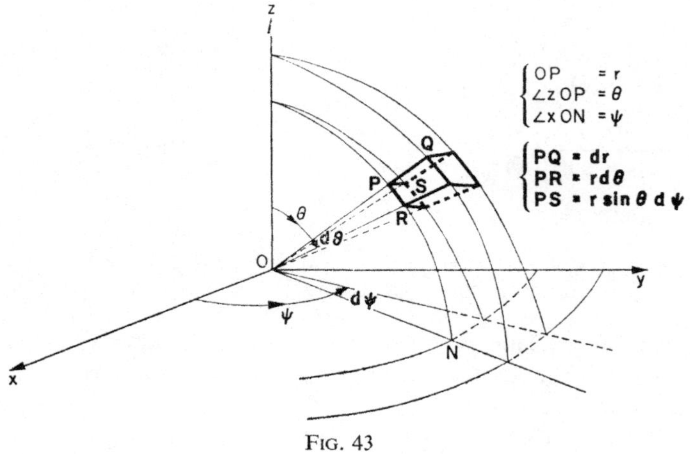

FIG. 43

to draw this figure for himself, and letter the remaining vertices of the volume element. It is immediately clear then that, for example, PR and QS' are not exactly equal; for $PR = r\,d\theta$, $QS' = (r + dr)\,d\theta$. Thus PR and QS' are equal only to the first order of small quantities. It is also clear that PQ, RS' are equal, each being of length dr.

We shall write $\mathbf{V} = V_r\hat{\mathbf{r}} + V_\theta\hat{\boldsymbol{\theta}} + V_\psi\hat{\boldsymbol{\psi}}$. Then using the results of § 7:7 with the appropriate values for h_1, h_2, h_3, V_1, V_2, V_3, we obtain the expressions for div, grad and curl in this system. These are

$$\operatorname{grad}\phi = \frac{\partial\phi}{\partial r}\hat{\mathbf{r}} + \frac{1}{r}\frac{\partial\phi}{\partial\theta}\hat{\boldsymbol{\theta}} + \frac{1}{r\sin\theta}\frac{\partial\phi}{\partial\psi}\hat{\boldsymbol{\psi}}, \tag{5}$$

$$\operatorname{div}\mathbf{V} = \frac{1}{r^2\sin\theta}\left[\frac{\partial}{\partial r}(r^2\sin\theta\, V_r) + \frac{\partial}{\partial\theta}(r\sin\theta\, V_\theta) + \frac{\partial}{\partial\psi}(r\, V_\psi)\right], \tag{6}$$

$$\operatorname{curl}\mathbf{V} = \frac{1}{r\sin\theta}\left[\frac{\partial}{\partial\theta}(V_\psi\sin\theta) - \frac{\partial V_\theta}{\partial\psi}\right]\hat{\mathbf{r}} + $$
$$+\left[\frac{1}{r\sin\theta}\frac{\partial V_r}{\partial\psi} - \frac{1}{r}\frac{\partial}{\partial r}(rV_\psi)\right]\hat{\boldsymbol{\theta}} + \frac{1}{r}\left[\frac{\partial}{\partial r}(rV_\theta) - \frac{\partial V_r}{\partial\theta}\right]\hat{\boldsymbol{\psi}}. \tag{7}$$

Also since $\nabla^2\phi = \operatorname{div}\operatorname{grad}\phi$, then taking $\mathbf{V} = \operatorname{grad}\phi$ and applying (6),

$$\nabla^2\phi = \frac{1}{r^2\sin\theta}\left[\frac{\partial}{\partial r}\left(r^2\sin\theta\frac{\partial\phi}{\partial r}\right) + \frac{\partial}{\partial\theta}\left(\sin\theta\frac{\partial\phi}{\partial\theta}\right) + \frac{\partial}{\partial\psi}\left(\frac{1}{\sin\theta}\frac{\partial\phi}{\partial\psi}\right)\right].$$

This reduces to

$$\nabla^2\phi = \frac{\partial^2\phi}{\partial r^2} + \frac{1}{r^2}\frac{\partial^2\phi}{\partial\theta^2} + \frac{1}{r^2\sin^2\theta}\frac{\partial^2\phi}{\partial\psi^2} + \frac{2}{r}\frac{\partial\phi}{\partial r} + \frac{\cot\theta}{r}\frac{\partial\phi}{\partial\theta}, \tag{8}$$

which was the expression obtained in 2:4(13) by another method.

By resolving the unit vectors $\hat{\mathbf{r}}$, $\hat{\boldsymbol{\theta}}$, $\hat{\boldsymbol{\psi}}$ in the directions of $\mathbf{i}$, $\mathbf{j}$, $\mathbf{k}$, we obtain

$$\hat{\mathbf{r}} = \cos\psi\sin\theta\,\mathbf{i} + \sin\psi\sin\theta\,\mathbf{j} + \cos\theta\,\mathbf{k}, \tag{9a}$$

$$\hat{\boldsymbol{\theta}} = \cos\psi\cos\theta\,\mathbf{i} + \sin\psi\cos\theta\,\mathbf{j} - \sin\theta\,\mathbf{k}, \tag{9b}$$

$$\hat{\boldsymbol{\psi}} = -\sin\psi\,\mathbf{i} + \cos\psi\,\mathbf{j}. \tag{9c}$$

Equations (9) show clearly how the unit vectors $\hat{\mathbf{r}}$, $\hat{\boldsymbol{\theta}}$, $\hat{\boldsymbol{\psi}}$ depend on the variables θ, ψ; as θ, ψ change, so the directions of $\hat{\mathbf{r}}$, $\hat{\boldsymbol{\theta}}$, $\hat{\boldsymbol{\psi}}$ change in accordance with equations (9). (In the general case the vectors $\hat{\mathbf{a}}$, $\hat{\mathbf{b}}$, $\hat{\mathbf{c}}$ have components which are functions of u, v and w; in the particular case of spherical polars only θ, ψ of the three variables r, θ, ψ are involved.)

(ii) *Cylindrical polar coordinates*

We transform from x, y, z, to cylindrical polar coordinates ρ, ψ, z by means of the equations

$$x = \rho\cos\psi, \quad y = \rho\sin\psi, \quad z = z, \tag{10}$$

where $$\rho \geqq 0, \quad 0 \leqq \psi < 2\,.$$

Thus in Fig. 44 the distance ON is equal to ρ, N being the foot of

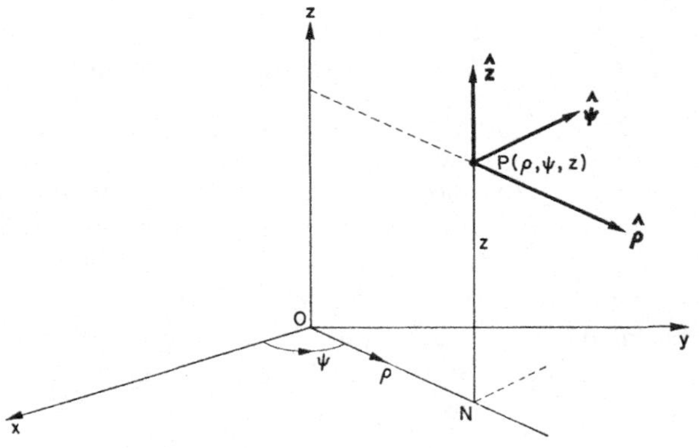

FIG. 44

the perpendicular from P to the x–y plane, while ψ lies in the x–y plane and is measured from the x-axis towards ON, and the third coordinate z is that of rectangular cartesian coordinates which is unchanged by the transformation.

The coefficients g_{ij} are easily calculated; we find

$$g_{11} = 1, \quad g_{22} = \rho^2, \quad g_{33} = 1, \quad g_{12} = g_{23} = g_{31} = 0.$$

Hence the line-element $\mathrm{d}l$ is given by

$$\mathrm{d}l^2 = \mathrm{d}\rho^2 + \rho^2\,\mathrm{d}\psi^2 + \mathrm{d}z^2, \tag{11}$$

and, of course,

$$h_1 = 1, \quad h_2 = \rho, \quad h_3 = 1. \tag{12}$$

The element of volume is

$$h_1 h_2 h_3\,\mathrm{d}u\,\mathrm{d}v\,\mathrm{d}w = \rho\,\mathrm{d}\rho\,\mathrm{d}\psi\,\mathrm{d}z. \tag{13}$$

The coordinate surfaces follow easily from (10) or from Fig. 44. Since $x^2 + y^2 = \rho^2$, the surfaces $\rho =$ constant are right circular cylinders with axis the z-axis; $\psi =$ constant gives planes through the z-axis just as for spherical polars, and $z =$ constant gives planes parallel to the x–y plane.

At the point $P(\rho, \psi, z)$, we take unit vectors $\hat{\boldsymbol{\rho}}$, $\hat{\boldsymbol{\psi}}$, $\hat{\mathbf{z}}$, in the directions of ρ, ψ, z increasing. With the usual notation we find

$$PQ = \mathrm{d}s_1 = \mathrm{d}\rho, \quad PR = \mathrm{d}s_2 = \rho\,\mathrm{d}\psi, \quad PS = \mathrm{d}s_3 = \mathrm{d}z.$$

The volume element is shown in Fig. 45. The reader should draw this diagram and complete the lettering. As in the spherical polar system, we see that PR and QS' are equal only to the first order of small quantities; in this system, however, not only are PQ, RS' equal, but also PS, QR' are equal, each being of length $\mathrm{d}z$.

Writing $\mathbf{V} = V_\rho\hat{\boldsymbol{\rho}} + V_\psi\hat{\boldsymbol{\psi}} + V_z\hat{\mathbf{z}}$, we obtain

$$\operatorname{grad}\phi = \frac{\partial\phi}{\partial\rho}\hat{\boldsymbol{\rho}} + \frac{1}{\rho}\frac{\partial\phi}{\partial\psi}\hat{\boldsymbol{\psi}} + \frac{\partial\phi}{\partial z}\hat{\mathbf{z}}, \tag{14}$$

$$\operatorname{div}\mathbf{V} = \frac{1}{\rho}\left[\frac{\partial}{\partial\rho}(\rho V_\rho) + \frac{\partial V_\psi}{\partial\psi} + \frac{\partial}{\partial z}(\rho V_z)\right], \tag{15}$$

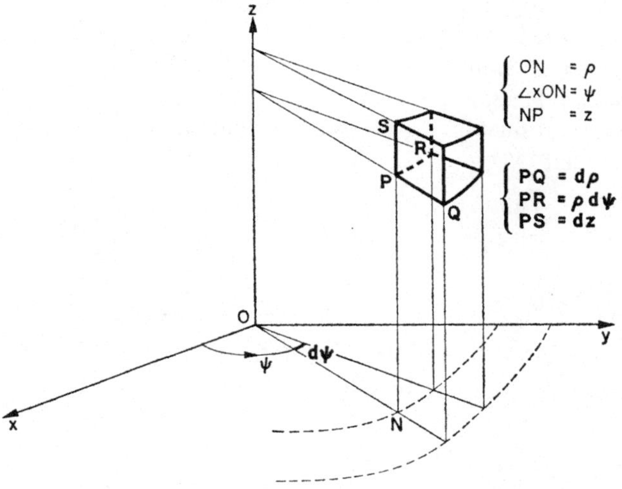

FIG. 45

$$\operatorname{curl} \mathbf{V} = \frac{1}{\rho}\left[\frac{\partial V_z}{\partial \psi} - \frac{\partial}{\partial z}(\rho V_\psi)\right]\hat{\boldsymbol{\rho}} + \left[\frac{\partial V_\rho}{\partial z} - \frac{\partial V_z}{\partial \rho}\right]\hat{\boldsymbol{\psi}} + $$

$$+ \frac{1}{\rho}\left[\frac{\partial}{\partial \rho}(\rho V_\psi) - \frac{\partial V_\rho}{\partial \psi}\right]\hat{\mathbf{z}}, \tag{16}$$

$$\nabla^2 \phi = \frac{1}{\rho}\left[\frac{\partial}{\partial \rho}\left(\rho\frac{\partial \phi}{\partial \rho}\right) + \frac{\partial}{\partial \psi}\left(\frac{1}{\rho}\frac{\partial \phi}{\partial \psi}\right) + \frac{\partial}{\partial z}\left(\rho\frac{\partial \phi}{\partial z}\right)\right].$$

The last result reduces to

$$\nabla^2 \phi = \frac{\partial^2 \phi}{\partial \rho^2} + \frac{1}{\rho^2}\frac{\partial^2 \phi}{\partial \psi^2} + \frac{\partial^2 \phi}{\partial z^2} + \frac{1}{\rho}\frac{\partial \phi}{\partial \rho}, \tag{17}$$

which is the form quoted in 2:4(8), apart from the term $\partial^2\phi/\partial z^2$.

Resolving the unit vectors $\hat{\boldsymbol{\rho}}$, $\hat{\boldsymbol{\psi}}$, $\hat{\mathbf{z}}$ in the direction of **i**, **j**, **k** we obtain

$$\hat{\boldsymbol{\rho}} = \cos\psi\,\mathbf{i} + \sin\psi\,\mathbf{j}, \tag{18a}$$

$$\hat{\boldsymbol{\psi}} = -\sin\psi\,\mathbf{i} + \cos\psi\,\mathbf{j}, \tag{18b}$$

$$\hat{\mathbf{z}} = \mathbf{k}. \tag{18c}$$

We give now three examples using cylindrical and spherical polar coordinates.

EXAMPLE 1. *Show that the vector field* $\mathbf{V} = (z/\rho)\hat{\boldsymbol{\psi}} + \psi\hat{\mathbf{z}}$ *is both irrotational and solenoidal.*

$\mathbf{V}$ is irrotational if curl $\mathbf{V} = \mathbf{0}$. In cylindrical polars, using (16), we find that

$$\text{curl}\,\mathbf{V} = \frac{1}{\rho}\left[\frac{\partial}{\partial\psi}(\psi) - \frac{\partial}{\partial z}(z)\right]\hat{\boldsymbol{\rho}} + \left[-\frac{\partial\psi}{\partial\rho}\right]\hat{\boldsymbol{\psi}} + \frac{1}{\rho}\left[\frac{\partial z}{\partial\rho}\right]\hat{\mathbf{z}}$$

$$= \mathbf{0}, \text{ since } d\psi/\partial\rho = \partial z/\partial\rho = 0.$$

Alternatively, $\mathbf{V}$ is irrotational if there exists a scalar field ϕ such that grad $\phi = \mathbf{V}$. In cylindrical polars

$$\text{grad}\,\phi = \frac{\partial\phi}{\partial\rho}\hat{\boldsymbol{\rho}} + \frac{1}{\rho}\frac{\partial\phi}{\partial\psi}\hat{\boldsymbol{\psi}} + \frac{\partial\phi}{\partial z}\hat{\mathbf{z}}.$$

So here we require

$$\frac{1}{\rho}\frac{\partial\phi}{\partial\psi} = \frac{z}{\rho}, \quad \frac{\partial\phi}{\partial z} = \psi, \quad \frac{\partial\phi}{\partial\rho} = 0.$$

These equations are satisfied by $\phi = z\psi$ + constant. So a scalar potential function ϕ exists and hence $\mathbf{V}$ is irrotational.

Further, $\mathbf{V}$ is solenoidal if div $\mathbf{V} = 0$; using (15) we find here that

$$\text{div}\,\mathbf{V} = \frac{1}{\rho}\left[\frac{\partial}{\partial\psi}\left(\frac{z}{\rho}\right) + \frac{\partial}{\partial z}\Big(\rho\psi\Big)\right] = 0.$$

Hence $\mathbf{V}$ is solenoidal. Alternatively, since we have shown that $\mathbf{V}$ = grad ϕ, where $\phi = z\psi$ + constant, $\mathbf{V}$ is solenoidal if div grad $\phi = 0$. Using (17) to obtain $\nabla^2\phi$ in this case we have at once that $\nabla^2\phi = 0$, since $\partial\phi/\partial\rho = 0$, $\partial\phi/\partial\psi = z$, $\partial\phi/\partial z = 0$.

EXAMPLE 2. *Using spherical polar coordinates find* div $(r^2\mathbf{r})$, *where* $\mathbf{r}$ *is the position vector of the point* $P(r, \theta, \psi)$.

In the spherical polar system $\mathbf{r} = r\hat{\mathbf{r}}$. Write $\mathbf{V} = r^2\mathbf{r} = r^3\hat{\mathbf{r}}$.

Then, using (15) with $V_r = r^3$, $V_\theta = V_\psi = 0$,

$$\operatorname{div} \mathbf{V} = \frac{1}{r^2 \sin\theta}\left[\frac{\partial}{\partial r}(r^2 \sin\theta \,.\, r^3)\right] = \frac{1}{r^2 \sin\theta} \,.\, 5r^4 \sin\theta,$$

so
$$\operatorname{div}(r^2\mathbf{r}) = 5r^2.$$

(See §§ 6:7, 6:8, for solutions of this example in rectangular cartesian coordinates.)

EXAMPLE 3. *Show that the unit vector $\hat{\boldsymbol{\psi}}$ of the spherical polar system is solenoidal but not irrotational.*

Writing $\mathbf{V} = \hat{\boldsymbol{\psi}}$, we have $V_r = V_\theta = 0$, $V_\psi = 1$. Hence, using (15), div $\mathbf{V} = 0$; that is, $\mathbf{V} = \hat{\boldsymbol{\psi}}$ is solenoidal. Using (16),

$$\operatorname{curl} \mathbf{V} = \frac{1}{r \sin\theta}\left[\frac{\partial}{\partial\theta}(\sin\theta)\right]\hat{\mathbf{r}} + \left[-\frac{1}{r}\frac{\partial}{\partial r}(r)\right]\hat{\boldsymbol{\theta}} \neq \mathbf{0}.$$

Hence $\mathbf{V} = \hat{\boldsymbol{\psi}}$ is not irrotational.

Exercises 7:8

1. The constants c_1, c_2, c_3 are chosen so that the surfaces

$$x^2 + y^2 + z^2 = c_1^2, \quad x^2 + y^2 = c_2^2 z^2, \quad y = c_3 x$$

all pass through the point $P(x_1, y_1, z_1)$. Obtain expressions in terms of x_1, y_1, z_1, **i**, **j**, **k** for vectors normal to these surfaces at P. Hence deduce the expressions given in § 7:8 (i) for the unit vectors $\hat{\mathbf{r}}$, $\hat{\boldsymbol{\theta}}$, $\hat{\boldsymbol{\psi}}$ of the spherical polar coordinate system in terms of θ, ψ, **i**, **j**, **k**.

2. The paraboloidal coordinate system is given by the equations

$$x = \xi\eta \cos\psi, \quad y = \xi\eta \sin\psi, \quad z = \tfrac{1}{2}(\xi^2 - \eta^2).$$

Find the curvilinear components of grad ϕ in this system, where $\phi = \log r$. Also, if f is any scalar point function, find the expression for $\nabla^2 f$ in this system.

3. Elliptic cylindrical coordinates u, v, z are given by the equations

$$x = k \cosh u \cos v, \quad y = k \sinh u \sin v, \quad z = z,$$

where k is a constant.

(i) Find the form of the coordinate surfaces and show that they constitute an orthogonal system.

(ii) Find expressions for the line-element dl, the volume element, $\nabla\phi$ and $\nabla^2\phi$ in this system, where ϕ is any scalar field.

(iii) Find expressions for the unit vectors $\hat{\mathbf{u}}$, $\hat{\mathbf{v}}$, $\hat{\mathbf{z}}$ at the point (u, v, z) in terms of **i**, **j**, **k**, and check that they are mutually orthogonal.

4. Parabolic cylindrical coordinates u, v, z are given by the equations $x = \frac{1}{2}(u^2 - v^2)$, $y = uv$, $z = z$. Find the form of the coordinate surfaces and show that they constitute an orthogonal system. Find also expressions for the unit vectors $\hat{\mathbf{u}}$, $\hat{\mathbf{v}}$, $\hat{\mathbf{z}}$ of this system in terms of $\mathbf{i}$, $\mathbf{j}$, $\mathbf{k}$; check that $\hat{\mathbf{u}}$, $\hat{\mathbf{v}}$, $\hat{\mathbf{z}}$ are mutually orthogonal.
5. A vector field $\mathbf{V}$ is given in the spherical polar system by the equation

$$\mathbf{V} = r \sin \theta \hat{\mathbf{r}} + r \cos \theta \hat{\boldsymbol{\theta}} - r(1 + \sin^2 \theta)\psi \hat{\boldsymbol{\psi}}.$$

Show that this vector field is solenoidal but not irrotational.

Miscellaneous Exercises 7

(Throughout these examples, the position vector $x\mathbf{i} + y\mathbf{j} + z\mathbf{k}$ of the point (x, y, z) is denoted by $\mathbf{r}$.)

1. (i) If $\mathbf{A}$, $\mathbf{B}$ are any two vector point functions, establish the identities
   (a) $\operatorname{div} (\mathbf{A} \times \mathbf{B}) = \mathbf{B} \cdot \operatorname{curl} \mathbf{A} - \mathbf{A} \cdot \operatorname{curl} \mathbf{B}$,
   (b) $\operatorname{curl} (\mathbf{A} \times \mathbf{B}) = (\operatorname{div} \mathbf{B})\mathbf{A} - (\operatorname{div} \mathbf{A})\mathbf{B} + (\mathbf{B} \cdot \nabla)\mathbf{A} - (\mathbf{A} \cdot \nabla)\mathbf{B}$.
   Evaluate $\operatorname{div} (\mathbf{A} \times \mathbf{B})$, $\operatorname{curl} (\mathbf{A} \times \mathbf{B})$ using these identities, when

$$\mathbf{A} = x\mathbf{i} + 2y\mathbf{j} + z\mathbf{k}, \quad \mathbf{B} = 2x\mathbf{i} - y\mathbf{j} + 2z\mathbf{k}.$$

   Check your results by another method.
   (ii) If $\mathbf{A}$ and $\mathbf{B}$ are such that $\mathbf{A} \cdot \operatorname{curl} \mathbf{B} = \mathbf{B} \cdot \operatorname{curl} \mathbf{A}$, show that $\mathbf{A} \times \mathbf{B}$ is solenoidal, and that it satisfies the equation

$$\operatorname{curl} \operatorname{curl} \mathbf{V} + \nabla^2 \mathbf{V} = \mathbf{0}.$$

2. If the vector field $\mathbf{V}$ is given by the equation $\mathbf{V} = \operatorname{curl} (\phi \mathbf{k})$, where ϕ is a scalar point function of x and y only, show that

$$\operatorname{curl} (\mathbf{V} \cdot \nabla)\mathbf{V} = (\mathbf{V} \cdot \nabla) \operatorname{curl} \mathbf{V} = \operatorname{grad} \phi \times \operatorname{grad} \nabla^2 \phi.$$

3. If the vector field $\mathbf{V}$ is given by the equation $\mathbf{V} = \nabla(z\phi) + \alpha\phi\mathbf{k}$, where α is a constant and ϕ is a scalar point function of x, y, z such that $\nabla^2 \phi = 0$, show that $\operatorname{div} \mathbf{V} = (2 + \alpha)\, \partial\phi/\partial z$. Using the identity

$$\operatorname{curl} \operatorname{curl} \mathbf{A} \equiv \operatorname{grad} \operatorname{div} \mathbf{A} - \nabla^2 \mathbf{A},$$

   or otherwise, show that

$$\nabla^2 \mathbf{V} = \nabla\left(2\frac{\partial \phi}{\partial z}\right).$$

   Deduce that $\mathbf{V}$ is a solution of the equation

$$\operatorname{grad} \operatorname{div} \mathbf{V} + (1 - 2\eta)\nabla^2 \mathbf{V} = \mathbf{0}$$

   if $\alpha = 4(\eta - 1)$. [B.Sc.(Eng.)]

4. Prove that div $\mathbf{B} = 0$ is the necessary and sufficient condition that a given vector $\mathbf{B}$ can be expressed in the form $\mathbf{B} = \text{curl}\,\mathbf{A}$.

(i) Find $\mathbf{A} = A_1\mathbf{i} + A_2\mathbf{j} + A_3\mathbf{k}$, given that $A_3 = 0$ and

$$\mathbf{B} = r^{-5}(-yz\mathbf{i} - xz\mathbf{j} + 2xy\mathbf{k}).$$

(ii) If $\mathbf{B} = \text{grad}\,\phi$, where $\phi = \mathbf{M}\,.\,\text{grad}\,(1/r)$, $\mathbf{M}$ being a constant vector, verify that $\mathbf{A} = \text{grad}\,(1/r) \times \mathbf{M}$ is a solution for $\mathbf{A}$ which also satisfies the equation div $\mathbf{A} = 0$. [B.Sc.(Spec.)]

5. The vector $\mathbf{B}$ is defined at the point $P(x, y, z)$ in the form of a line integral along a curve C (not passing through P), where

$$\mathbf{B} = \oint_C \frac{1}{r}\,\mathrm{d}\mathbf{s},$$

$\mathrm{d}\mathbf{s}$ being a vector line element at a point Q of C, and $\overrightarrow{QP} = \mathbf{r}$. Show that

$$\text{curl}\,\mathbf{B} = \oint_C \frac{\mathrm{d}\mathbf{s} \times \mathbf{r}}{r^3},$$

and evaluate this integral when C is the curve $x^2 + y^2 = a^2$, $z = 0$, and P is the point $(0, 0, a)$. [B.Sc.(Eng.)]

6. Starting with the Divergence Theorem $\int_S \mathbf{A}\,.\,\mathrm{d}\mathbf{S} = \int_V \text{div}\,\mathbf{A}\,\mathrm{d}v$, show by taking $\mathbf{A} = \mathbf{m} \times \mathbf{B}$, where $\mathbf{m}$ is an arbitrary constant vector, that

$$\int_S \hat{\mathbf{n}} \times \mathbf{B}\,\mathrm{d}S = \int_V \text{curl}\,\mathbf{B}\,\mathrm{d}v,$$

S being a simple closed surface enclosing a volume V, and $\hat{\mathbf{n}}$ the unit vector along the outward normal to S at the point where $\mathrm{d}S$ is situated.

Given that $\phi = (x + 1)^2 + 2(y - 1)^2 + z^2$, find the value of

$$\int \hat{\mathbf{n}} \times \text{grad}\,\phi\,\mathrm{d}S,$$

taken over that part of the ellipsoidal surface $x^2/a^2 + y^2/b^2 + z^2/c^2 = 1$ for which $z \geqq 0$. [B.Sc.(Spec.)]

7. By considering the quantity div $(\phi\mathbf{c} \times \mathbf{u})$, where ϕ, $\mathbf{u}$ are scalar and vector point functions respectively, and $\mathbf{c}$ is an arbitrary constant vector, show that

$$\int_V \phi\,\text{curl}\,\mathbf{u}\,\mathrm{d}v = \int_V \mathbf{u} \times \text{grad}\,\phi\,\mathrm{d}v - \int_S \phi\mathbf{u} \times \mathrm{d}\mathbf{S},$$

the last integral being over a closed surface S and the other integrals being taken through the volume V enclosed by S.

8. Assuming the Divergence Theorem, prove that

$$\int_S \phi \mathbf{P} \cdot d\mathbf{S} = \int_V \mathbf{P} \cdot \operatorname{grad} \phi \, dv + \int_V \phi \operatorname{div} \mathbf{P} \, dv,$$

$$\int_S \mathbf{Q}(\mathbf{P} \cdot d\mathbf{S}) = \int_V (\mathbf{P} \cdot \nabla)\mathbf{Q} \, dv + \int_V \mathbf{Q} \operatorname{div} \mathbf{P} \, dv,$$

where ϕ is a scalar field and **P**, **Q** are vector fields. Hence prove that

$$\int_V (\mathbf{E} \times \mathbf{r}) \operatorname{div} \mathbf{E} \, dv$$

$$= \int_S (\mathbf{E} \times \mathbf{r})(\mathbf{E} \cdot d\mathbf{S}) + \tfrac{1}{2}\int_S E^2 \mathbf{r} \times d\mathbf{S} + \int_V (\mathbf{E} \times \operatorname{curl} \mathbf{E}) \times \mathbf{r} \, dv,$$

where **E** is a vector field. [B.Sc.(Spec.)]

9. Show that if ϕ, $\boldsymbol{\omega}$ are scalar and vector fields respectively, such that $\operatorname{grad} \phi = \operatorname{curl} \boldsymbol{\omega}$, then ϕ satisfies Laplace's equation $\nabla^2\phi = 0$. Show also that if V is a volume bounded by a closed surface S then

$$\int_S (\boldsymbol{\omega} \times \operatorname{curl} \boldsymbol{\omega}) \cdot \hat{\mathbf{n}} \, dS = \int_V |\operatorname{grad} \phi|^2 dv,$$

where $\hat{\mathbf{n}}$ is a unit vector in the direction of the outward normal to S at the point where dS is situated. [B.Sc.(Eng.)]

10. If $\mathbf{A} = \operatorname{curl} \mathbf{B}$ and $\mathbf{B} = \operatorname{curl} \mathbf{C}$, and if **C** is a suitable vector field, show that

$$\int_V B^2 \, dv = \int_S (\mathbf{C} \times \mathbf{B}) \cdot \hat{\mathbf{n}} \, dS + \int_V \mathbf{C} \cdot \mathbf{A} \, dv,$$

where $\hat{\mathbf{n}}$ is the outward normal to a surface S enclosing a volume V.

If u, **H** are suitable scalar and vector fields respectively, prove that

$$\operatorname{curl}(u\mathbf{H}) = \nabla u \times \mathbf{H} + u \operatorname{curl} \mathbf{H}.$$

Hence, or otherwise, show that

$$\int_S u\hat{\mathbf{n}} \cdot \operatorname{curl} \mathbf{H} \, dS = \oint_C u\mathbf{H} \cdot d\mathbf{r} - \int_S (\nabla u \times \mathbf{H}) \cdot \hat{\mathbf{n}} \, dS,$$

where $\hat{\mathbf{n}}$ is the outward normal to an open surface S which has the closed curve C as boundary. By putting $\mathbf{H} = \nabla w$, or otherwise, show that

$$\oint_C u\nabla w \cdot d\mathbf{r} = -\oint_C w\nabla u \cdot d\mathbf{r},$$

where w is another scalar field. [B.Sc.(Gen.)]

11. A solid in the positive octant is bounded by the planes $x = 0$, $y = 0$, $z = 0$, and the cylinders $x^2 + y^2 = a^2$, $x^2 + z^2 = a^2$. If $\mathbf{V}$ is given by the equation

$$\mathbf{V} = 2yz\mathbf{i} + (x + 3y - 2)\mathbf{j} - (x^2 + z)\mathbf{k},$$

show that the value of the integral $\int(\nabla \times \mathbf{V}) \,.\, d\mathbf{S}$ taken over the curved surface of the solid is $\frac{1}{12}a^2(8a + 3\pi)$. [B.Sc.(Spec.)]

12. The components of a vector $\mathbf{V}$ with respect to spherical polar coordinates (r, θ, ψ) are (u, v, w); write down the expressions for the components of curl $\mathbf{V}$ in terms of u, v, w.

If $u = v = \partial w/\partial\psi = 0$, and curl curl $\mathbf{V} = \mathbf{0}$, show that

$$r^2\frac{\partial^2 w}{\partial r^2} + 2r\frac{\partial w}{\partial r} + \frac{\partial}{\partial\theta}\left(\frac{\partial w}{\partial\theta} + w\cot\theta\right) = 0.$$

Show also that there is a solution of this equation of the form $f(r)\sin\theta$, and find the general expression for $f(r)$. [B.Sc.(Spec.)]

13. Parabolic cylindrical coordinates (λ, μ, z) are defined by the equations $x = \frac{1}{2}(\lambda^2 - \mu^2)$, $y = \lambda\mu$, $z = z$. Find the form taken by Laplace's equation $\nabla^2\phi = 0$ in this system. Show that it has solutions of the form $L(\lambda)M(\mu)\cos az$, where $L(\lambda)$, $M(\mu)$ are respectively functions of λ and μ only, if $L(\lambda)$ satisfies the equation

$$\frac{d^2L}{d\lambda^2} - (a^2\lambda^2 + b)L = 0,$$

and if $M(\mu)$ satisfies a similar equation, a and b being constants. [B.Sc.(Gen.)]

14. Orthogonal curvilinear coordinates (u, v, ϕ) are given in terms of x, y, z by the equations

$$x = \mu(u, v)\cos\phi, \quad y = \mu(u, v)\sin\phi, \quad z = u\mu^4(u, v),$$

where $\mu(u, v)$ is a function of u and v satisfying the equation $4u^2\mu^8 + \mu^2 = v$. Determine the nature of the coordinate surfaces.

Show that in this coordinate system Laplace's equation $\nabla^2 f$ takes the form

$$\frac{\partial}{\partial u}\left(\mu^{-4}\frac{\partial f}{\partial u}\right) + 4\frac{\partial}{\partial v}\left(\mu^6\frac{\partial f}{\partial v}\right) + \frac{\mu^2}{H}\frac{\partial^2 f}{\partial\phi^2} = 0,$$

where $H = 1 + 16u^2\mu^6$. [B.Sc.(Spec.)]

15. Cylindrical polar coordinates (ρ, ψ, z) are given in terms of x, y, z by $x = \rho\cos\psi$, $y = \rho\sin\psi$, $z = z$. The unit vectors $\hat{\boldsymbol{\rho}}$, $\hat{\boldsymbol{\psi}}$, $\hat{\mathbf{z}}$ at the point (ρ, ψ, z) are respectively in the directions of ρ, ψ, z increasing. Write

down the expression for $\nabla\phi$ in terms of $\hat{\boldsymbol{\rho}}$, $\hat{\boldsymbol{\psi}}$, $\hat{\mathbf{z}}$, where ϕ is any scalar point function.

Show that

$$\frac{\partial}{\partial\psi}\hat{\boldsymbol{\rho}} = \hat{\boldsymbol{\psi}}, \quad \frac{\partial}{\partial\psi}\hat{\boldsymbol{\psi}} = -\hat{\boldsymbol{\rho}},$$

and hence obtain the values of $\nabla \cdot \mathbf{A}$ and $\nabla^2\phi$ where

$$\mathbf{A} = A_1\hat{\boldsymbol{\rho}} + A_2\hat{\boldsymbol{\psi}} + A_3\hat{\mathbf{z}}.$$

If χ is a homogeneous polynomial in ρ and z only of degree n, such that $\nabla^2\chi = 0$, and if $\phi = (\rho^2 + z^2)\chi$, show that $\nabla^2\phi = 2(2n + 3)\chi$.

[B.Sc.(Eng.)]

CHAPTER 8

APPLICATIONS OF VECTOR FIELD THEORY

8:1 Introduction

In this chapter we shall show how the basic ideas of vector field theory discussed in the last two chapters are applied in several branches of applied mathematics. The following pages are *not* intended as an introduction to the topics for readers completely unfamiliar with them; the aim is to link the theory of vector fields with the actual problems involved in each topic.

8:2 Hydrodynamics

(i) *The vector field* $\mathbf{q}$

We consider a non-viscous fluid in motion. Suppose that at time t a very small volume $\delta\tau$ of fluid is at the point P, where $\overrightarrow{OP} = \mathbf{r}$. If $\delta\tau$ is small enough, we can treat it as a particle. At time $t + \delta t$, this volume $\delta\tau$ will have moved to Q, where $\overrightarrow{OQ} = \mathbf{r} + \delta\mathbf{r}$ (see Fig. 46). The velocity $\mathbf{q}$ of the fluid is defined by

$$\mathbf{q} = \lim_{\delta t \to 0} \frac{\delta \mathbf{r}}{\delta t}. \tag{1}$$

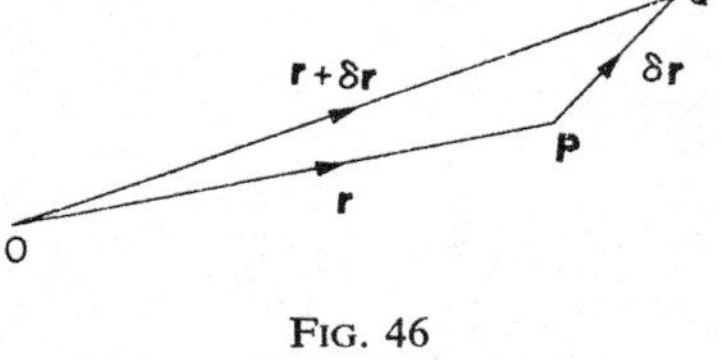

FIG. 46

In general, each particle in the moving volume of fluid will have at time t a velocity $\mathbf{q}$ which varies from point to point within the volume, and also varies with time. When $\mathbf{q}$ is independent of t, so that the velocity at any point P is the same at all times, then the motion is *steady*. Clearly the velocity vector $\mathbf{q}$ defines a *vector field*; at any given time t, $\mathbf{q}$ has a unique value at each point of the volume of fluid. We can write

$$\mathbf{q} = u\mathbf{i} + v\mathbf{j} + w\mathbf{k}, \tag{2}$$

where, in general, u, v, w are each functions of x, y, z, t; in steady motion, u, v, w are each functions of x, y and z only.

(ii) *Acceleration in a fluid*

We now consider the acceleration of the fluid particle $\delta\tau$. We have at the point P

$$\mathbf{q} = u\mathbf{i} + v\mathbf{j} + w\mathbf{k}, \tag{3}$$

and the velocity at Q is

$$\mathbf{q} + \delta\mathbf{q} = (u + \delta u)\mathbf{i} + (v + \delta v)\mathbf{j} + (w + \delta w)\mathbf{k}. \tag{4}$$

If P is the point (x, y, z), then Q is $(x + \delta x, y + \delta y, z + \delta z)$, where

$$\delta x = u\delta t, \quad \delta y = v\delta t, \quad \delta z = w\delta t. \tag{5}$$

To a first approximation, we have

$$\begin{aligned}\delta u &= \frac{\partial u}{\partial x}\delta x + \frac{\partial u}{\partial y}\delta y + \frac{\partial u}{\partial z}\delta z + \frac{\partial u}{\partial t}\delta t \\ &= \left(u\frac{\partial u}{\partial x} + v\frac{\partial u}{\partial y} + w\frac{\partial u}{\partial z} + \frac{\partial u}{\partial t}\right)\delta t.\end{aligned}$$

This can be expressed as

$$\delta u = \left(\mathbf{q} \,.\, \nabla u + \frac{\partial u}{\partial t}\right)\delta t, \tag{6}$$

with similar expressions for δv, δw. Then from (6), (3) and (4),

$$\delta \mathbf{q} = \left[\left(\mathbf{q} \cdot \nabla u + \frac{\partial u}{\partial t}\right)\mathbf{i} + \left(\mathbf{q} \cdot \nabla v + \frac{\partial v}{\partial t}\right)\mathbf{j} + \left(\mathbf{q} \cdot \nabla w + \frac{\partial w}{\partial t}\right)\mathbf{k}\right]\delta t. \tag{7}$$

We can rewrite this more simply as

$$\delta \mathbf{q} = \left[\frac{\partial \mathbf{q}}{\partial t} + (\mathbf{q} \cdot \nabla)\mathbf{q}\right]\delta t, \tag{8}$$

since $\partial/\partial t$, $\mathbf{q} \cdot \nabla$ are both scalar operators, and so act on each component of the vector $\mathbf{q}$, producing the result given in (7).

The acceleration of the fluid particle $\delta\tau$ is defined by

$$\lim_{\delta t \to 0} \frac{\delta \mathbf{q}}{\delta t} = \frac{\partial \mathbf{q}}{\partial t} + (\mathbf{q} \cdot \nabla)\mathbf{q}. \tag{9}$$

We can here write

$$\frac{\mathrm{d}}{\mathrm{d}t} \equiv \frac{\partial}{\partial t} + \mathbf{q} \cdot \nabla, \tag{10}$$

where on the left of (10), $\mathrm{d}/\mathrm{d}t$ implies "differentiation following the motion of the fluid", both time and position being taken into account. Sometimes the notation $\mathrm{D}/\mathrm{D}t$ is used, instead of $\mathrm{d}/\mathrm{d}t$. The operator in (10) can, of course, be applied to any scalar or vector functions connected with a fluid to give the rate of change as the fluid moves. In § 8:2 (iii) we shall apply it to the density, which is a scalar function.

(iii) *The equation of continuity*

Suppose we are concerned with a fluid whose density is $m(x, y, z, t)$ at the point (x, y, z); that is to say we assume m to vary from point to point and with time. Imagine a closed surface S in the fluid, enclosing a volume τ of fluid of mass M. We assume that *there is no source of supply of fluid within S*. The actual rate of flow of the fluid *out* through the surface S is simply $\int_S m\mathbf{q} \cdot \mathrm{d}\mathbf{S}$;

for if dS is the element of surface at any point on S, with the corresponding normal $\hat{\mathbf{n}}$, as in Fig. 47, then

$$m\mathbf{q} \,.\, \mathrm{d}\mathbf{S} = m\mathbf{q} \,.\, \mathrm{d}S\hat{\mathbf{n}} = m\;\mathrm{d}S\;\mathrm{q}_n,$$

where q_n is the component of $\mathbf{q}$ in the direction of $\hat{\mathbf{n}}$. Thus $m\;\mathrm{d}S\,q_n$ measures the actual rate of flow of the fluid through the element

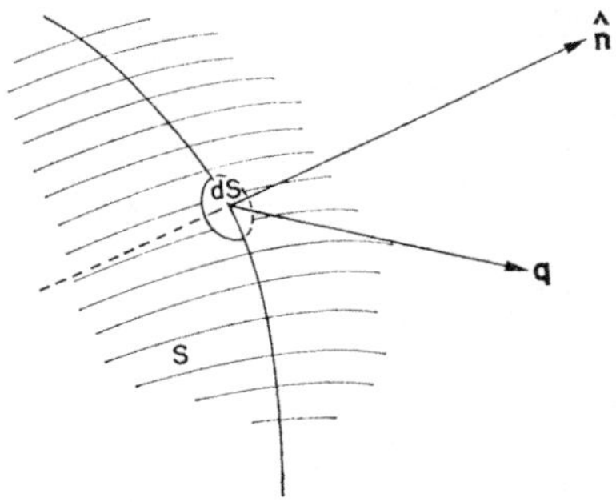

FIG. 47

of surface dS, and by integration over S we find the rate for the whole surface. Thus the *flux* of $m\mathbf{q}$ through the surface S (as defined in § 6:6) measures the rate at which fluid is flowing out across S, and hence measures the rate of decrease of mass within S.

If we now apply the Divergence Theorem [6:10(1)], we find that

$$\int_S (m\mathbf{q}) \,.\, \mathrm{d}\mathbf{S} = \int_\tau \operatorname{div}(m\mathbf{q})\,\mathrm{d}\tau. \tag{11}$$

Also the mass M within S is given by

$$M = \int_\tau m\,\mathrm{d}\tau$$

and so the rate of decrease of mass within S is

$$-\frac{\partial M}{\partial t} = -\frac{\partial}{\partial t}\int_\tau m\,\mathrm{d}\tau = -\int_\tau \frac{\partial m}{\partial t}\,\mathrm{d}\tau. \tag{12}$$

The rates given by (11) and (12) must be equal; so

$$-\int_\tau \frac{\partial m}{\partial t}\,\mathrm{d}\tau = \int_\tau \operatorname{div}(m\mathbf{q})\,\mathrm{d}\tau,$$

or

$$\int_\tau \left[\frac{\partial m}{\partial t} + \operatorname{div}(m\mathbf{q})\right]\mathrm{d}\tau = 0.$$

This is true for every volume τ we like to choose; it follows that at any point in a fluid the *equation of continuity*

$$\frac{\partial m}{\partial t} + \operatorname{div}(m\mathbf{q}) = 0 \tag{13}$$

is satisfied. All that we are saying here in mathematical language is that we cannot have any gaps within a volume of non-viscous fluid.

Using 6:8(3), we can rewrite (13) as

$$\frac{\partial m}{\partial t} + m\boldsymbol{\nabla}\,.\,\mathbf{q} + \mathbf{q}\,.\,\boldsymbol{\nabla}m = 0,$$

or

$$\frac{\mathrm{d}m}{\mathrm{d}t} + m\operatorname{div}\mathbf{q} = 0, \tag{14}$$

using the notation of § 8:2 (ii).

For an incompressible fluid, such as a liquid, m is constant and the equation of continuity (14) reduces to div $\mathbf{q} = 0$.

(iv) *Sources and sinks in incompressible fluid motion*

Suppose we now consider an incompressible fluid in motion in three dimensions, with a source of fluid at the origin. We can best visualise this by thinking of a pipe of very small diameter, with its open end at the origin, discharging fluid at a given rate of $4\pi ms$ units of volume in unit time, the fluid flowing out radially about the origin. An isolated point-source of fluid such as this does *not* occur in nature, but we need to study this idealised case to develop the theory of fluid flow. If we now consider the flow of fluid out across the surface of a sphere S of radius a and centre the origin, then, because the fluid is incompressible, $4\pi s$ units of volume must flow out of S in unit time. So we have

$$\int_S m\mathbf{q} \cdot d\mathbf{S} = 4\pi m s. \tag{15}$$

This relation does not involve a, the radius of S; the volume of fluid flowing out through all spheres centred at the origin in unit time is the same, because the fluid is incompressible.

We say here that we have an *isolated point-source of strength* s at the origin.

If we now imagine the flow reversed, so that fluid is flowing in through S and out at the origin, then we have a *sink of strength* s at the origin. This can be thought of as a source of strength $-s$.

A *uniform line-source* is obtained when we have point-sources distributed uniformly along a line.

An important case arises when we have a source of strength s at a point A, and a sink of strength s at a point B. If we let A move along AB towards B in such a way that sAB remains constant and equal to μ, say, then we have a *doublet* of strength μ with axis along the direction of BA.

It may happen that we have a fluid in motion in such a way that only two of the three variables x, y, z enter into the expression for $\mathbf{q}$. If we assume z to be absent, then it is sufficient to consider motion in the x–y plane only, and we have *two-dimensional flow*. Here we can also have a point-source of strength s at the origin; we suppose that it emits fluid sufficient to cover $2\pi s$ units of area in unit time. Then if we take a circle C of radius a and centre the origin, $2\pi s$ units of fluid area flow across the circle in unit time. If we consider the whole volume of fluid, then since motion is the same in all planes parallel to the x–y plane, we have a line-source along the z-axis. Reversing the direction of the flow, we would obtain a sink at the origin in the x–y plane, or in three dimensions a sink all along the z-axis. Our earlier definition of a doublet also holds in two-dimensional flow, A, B now lying in the x–y plane. [The factors 4π, 2π used above are convenient ones when we come to consider the velocity potentials in § 8:2 (vi).]

We could also discuss the motion due to sources distributed throughout a volume or over a surface; these in fact arise naturally in the theory of gravitational attraction; see § 8:3 (ii).

(v) *Rotational and irrotational flow*

We shall now distinguish between rotational and irrotational motion of a fluid of constant density. To do this, we consider the motion of a small spherical element in the moving field, whose centre at time t is the point $P(x, y, z)$ with respect to fixed axes $Oxyz$. Let $P'(x + x', y + y', z + z')$ be any other point within the sphere, as in Fig. 48. If $\mathbf{q}$ is the velocity of P at time t, we write

$$\mathbf{q} = u\mathbf{i} + v\mathbf{j} + w\mathbf{k}.$$

If $\mathbf{q}'$ is the velocity at P' at time t, where $\mathbf{q}' = u'\mathbf{i} + v'\mathbf{j} + w'\mathbf{k}$,

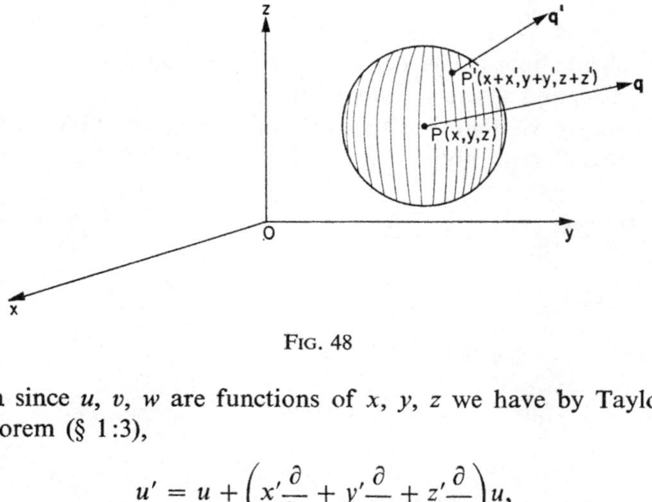

FIG. 48

then since u, v, w are functions of x, y, z we have by Taylor's Theorem (§ 1:3),

$$u' = u + \left(x'\frac{\partial}{\partial x} + y'\frac{\partial}{\partial y} + z'\frac{\partial}{\partial z}\right)u,$$

and two similar equations for v', w', each being correct to the first order in x', y', z'. Thus we can write

$$\mathbf{q}' = \mathbf{q} + \left(x'\frac{\partial \mathbf{q}}{\partial x} + y'\frac{\partial \mathbf{q}}{\partial y} + z'\frac{\partial \mathbf{q}}{\partial z}\right).$$

Suppose that m' is the mass of a fluid particle at P'; its linear momentum is $m'\mathbf{q}'$, and so, summing for the whole sphere, the linear momentum $\mathbf{L}$ is

$$\mathbf{L} = \sum(m'\mathbf{q}) + \sum m'\left(x'\frac{\partial \mathbf{q}}{\partial x} + y'\frac{\partial \mathbf{q}}{\partial y} + z'\frac{\partial \mathbf{q}}{\partial z}\right).$$

But $\mathbf{q}$, $\partial\mathbf{q}/\partial x$, $\partial\mathbf{q}/\partial y$, $\partial\mathbf{q}/\partial z$ are constant throughout the sphere at time t, since they refer to P, so that

$$\mathbf{L} = (\sum m')\mathbf{q} + (\sum m'x')\frac{\partial \mathbf{q}}{\partial x} + (\sum m'y')\frac{\partial \mathbf{q}}{\partial y} + (\sum m'z')\frac{\partial \mathbf{q}}{\partial z},$$

or $$\mathbf{L} = (\sum m')\mathbf{q}, \tag{16}$$

since $\Sigma m'x'$, $\Sigma m'y'$, $\Sigma m'z'$ are zero (they give the first moments of the sphere about planes parallel to the coordinate planes through its centre, and so are zero by symmetry).

Similarly, we find that the total angular momentum $\mathbf{A}$ of the spherical element about P is

$$\mathbf{A} = \sum(x'\mathbf{i} + y'\mathbf{j} + z'\mathbf{k}) \times m'\left(\mathbf{q} + x'\frac{\partial \mathbf{q}}{\partial x} + y'\frac{\partial \mathbf{q}}{\partial y} + z'\frac{\partial \mathbf{q}}{\partial z}\right)$$

$$= c_1\mathbf{i} + c_2\mathbf{j} + c_3\mathbf{k},$$

say, where on evaluation we find that

$$c_1 = \sum\left(m'y'w - m'z'v + m'x'y'\frac{\partial w}{\partial x} + m'y'^2\frac{\partial w}{\partial y} + m'y'z'\frac{\partial w}{\partial z} - \right.$$

$$\left. - m'z'x'\frac{\partial v}{\partial x} - m'y'z'\frac{\partial v}{\partial y} - mz'^2\frac{\partial v}{\partial x}\right).$$

Here again u, v, w and their partial derivatives are constant; also $\Sigma m'x'y'$, $\Sigma m'y'z'$, $\Sigma m'z'x'$ are all zero. So

$$c_1 = (\sum m'y'^2)\frac{\partial w}{\partial y} - (\sum m'z'^2)\frac{\partial v}{\partial z}.$$

Now for the spherical element

$$\sum m'x'^2 = \sum m'y'^2 = \sum m'z'^2 = \tfrac{1}{2}I,$$

where I is the moment of inertia about any diameter. So

$$c_1 = \tfrac{1}{2}I\left(\frac{\partial w}{\partial y} - \frac{\partial v}{\partial z}\right),$$

with similar expressions for c_2 and c_3. Thus finally

$$\begin{aligned}\mathbf{A} &= \tfrac{1}{2}I\left[\left(\frac{\partial w}{\partial y} - \frac{\partial v}{\partial z}\right)\mathbf{i} + \left(\frac{\partial u}{\partial z} - \frac{\partial w}{\partial x}\right)\mathbf{j} + \left(\frac{\partial v}{\partial x} - \frac{\partial u}{\partial y}\right)\mathbf{k}\right] \\ &= \tfrac{1}{2}I\,\mathrm{curl}\,\mathbf{q}. \qquad (17)\end{aligned}$$

Comparing equations (16) and (17) with the equations of motion of a solid sphere of the same radius and density, we find that if we could suddenly, at time t, solidify the fluid element, and remove the surrounding fluid, then P would move with velocity $\mathbf{q}$, and the sphere would rotate with angular velocity $\tfrac{1}{2}$ curl $\mathbf{q}$.

When the motion of the fluid is one of translation only, so that the spherical element is *not* rotating, then curl $\mathbf{q}$ is zero, and we say that the motion is *irrotational.* When curl $\mathbf{q} \neq \mathbf{0}$, the motion is *rotational*, and curl $\mathbf{q}$ is called the *vorticity* vector.

We saw in § 6:5 that an irrotational vector field $\mathbf{V}$ can always be expressed as the gradient of some scalar point function ϕ. It follows that in irrotational fluid motion there exists a *velocity potential function* ϕ such that

$$\mathbf{q} = -\nabla\phi, \qquad (18)$$

the negative sign being merely a convention in the definition of ϕ.

Some writers use $+\nabla\phi$ here. For an incompressible fluid, $\operatorname{div} \mathbf{q} = 0$; so in this case we obtain from (18)

$$\operatorname{div} \mathbf{q} = \nabla \cdot \nabla\phi = \nabla^2\phi = 0. \tag{19}$$

It follows at once that problems in non-viscous incompressible irrotational fluid motion require the solution of Laplace's equation under certain given boundary conditions (depending on the problem), and to get such solutions it is frequently necessary to transform Laplace's equation to a coordinate system other than the cartesian one.

It should be noticed that we can add to ϕ any arbitrary constant or any function of t without altering the value of $\mathbf{q}$; to this extent ϕ is always indeterminate.

We shall return to the velocity potential in the next section; now we obtain *Euler's equation of motion*. We consider the forces acting on the fluid within an imaginary closed surface S which moves with the fluid. Let $\mathbf{q}$ be the velocity of a fluid particle of volume $\mathrm{d}v$ and mass $m\,\mathrm{d}v$. Then the linear momentum of the particle is $m\,\mathrm{d}v\,\mathbf{q}$, and the rate of change of linear momentum is $m\,\mathrm{d}v(\mathrm{d}\mathbf{q}/\mathrm{d}t)$, in the notation of § 8:2 (ii). Suppose that there is an external force $\mathbf{F}$ per unit mass acting on the moving fluid; the force on the element $\mathrm{d}v$ is $m\,\mathrm{d}v\mathbf{F}$. If V is the volume contained by S, then the corresponding quantities for the total fluid V are

$$\int_V m\frac{\mathrm{d}\mathbf{q}}{\mathrm{d}t}\mathrm{d}v, \quad \int_V m\mathbf{F}\,\mathrm{d}v.$$

The only other force on the fluid in S is the pressure acting inwards across the surface S due to the surrounding fluid. If p is the fluid pressure per unit area, then on an element $\mathrm{d}S$ of S the force due to pressure is $p\,\mathrm{d}\mathbf{S}$ inwards; that is $-p\hat{\mathbf{n}}\,\mathrm{d}S$, where $\hat{\mathbf{n}}$ is the unit outward normal to $\mathrm{d}S$ (see Fig. 49). So for the whole surface the force due to pressure is

$$-\int_S p\hat{\mathbf{n}}\,\mathrm{d}S = -\int p\,\mathrm{d}\mathbf{S}.$$

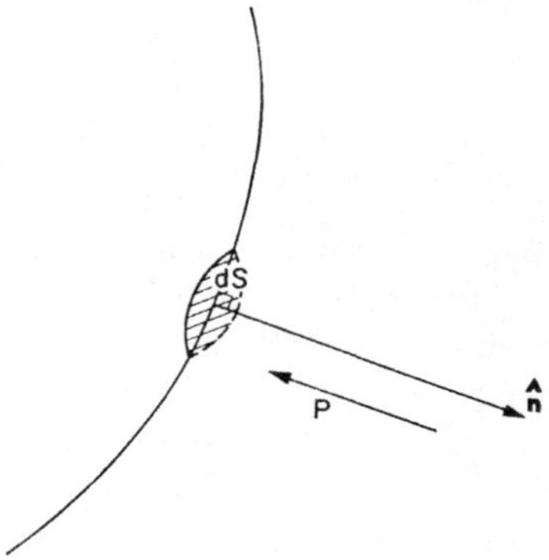

FIG. 49

Now

$$\int_S p\,\mathrm{d}\mathbf{S} = \int_V \nabla p\,\mathrm{d}v$$

(see Miscellaneous Exercises 6, No. 12).

We equate the rate of change of linear momentum

$$\int_V m\frac{\mathrm{d}\mathbf{q}}{\mathrm{d}t}\mathrm{d}v$$

to the external forces acting on the fluid within S which are $\int_V m\mathbf{F}\,\mathrm{d}v$ and $-\int_V \nabla p\,\mathrm{d}v$. So

$$\int_V m\frac{\mathrm{d}\mathbf{q}}{\mathrm{d}t}\mathrm{d}v = \int_V (m\mathbf{F} - \nabla p)\,\mathrm{d}v.$$

This is true for any closed surface in the fluid; hence

$$m\frac{\mathrm{d}\mathbf{q}}{\mathrm{d}t} = (m\mathbf{F} - \nabla p)$$

or

$$\frac{\mathrm{d}\mathbf{q}}{\mathrm{d}t} = \mathbf{F} - \frac{1}{m}\nabla p. \tag{20}$$

This is *Euler's equation of motion*. Using the alternative form for d/dt [8:2(10)], we can write (20) as

$$\frac{\partial \mathbf{q}}{\partial t} + (\mathbf{q} \,.\, \nabla)\mathbf{q} = \mathbf{F} - \frac{1}{m}\nabla p. \tag{21}$$

It frequently happens that $\mathbf{F}$ is a conservative force; the condition for this is curl $\mathbf{F} = \mathbf{0}$, so that there exists some scalar function K such that $\mathbf{F} = \nabla K$. We can further modify (6) by replacing $(\mathbf{q} \,.\, \nabla)\mathbf{q}$ by $\frac{1}{2}\nabla \mathbf{q}^2 - \mathbf{q} \times (\nabla \times \mathbf{q})$ (see Exercises 7:5, No. 3); then assuming that $\mathbf{F}$ is conservative, (21) becomes

$$\frac{\partial \mathbf{q}}{\partial t} + \tfrac{1}{2}\nabla \mathbf{q}^2 - \mathbf{q} \times (\nabla \times \mathbf{q}) = \nabla K - \frac{1}{m}\nabla p. \tag{22}$$

In the important case of irrotational fluid motion, $\nabla \times \mathbf{q} = \mathbf{0}$, and (22) reduces further.

(vi) *Some special irrotational velocity fields*

We shall now look briefly at three special potential functions which give important cases of irrotational flow of fluids of constant density. The reader will remember from the previous section that the equation of continuity implies that for motion of this type ϕ must satisfy Laplace's equation.

(a) $\phi = -kx$, *k a positive constant.* Clearly Laplace's equation is satisfied everywhere. Further since $\mathbf{q} = -\nabla\phi$, we have here

$$\mathbf{q} = k\mathbf{i},$$

giving steady motion at speed k in the direction of the x-axis (the motion is steady since $\mathbf{q}$ is independent of t). This type of motion is called uniform streaming.

We could take $\phi = k_1x + k_2y$, in which case

$$\mathbf{q} = k_1\mathbf{i} + k_2\mathbf{j},$$

so that streaming occurs with speed $\sqrt{(k_1^2 + k_2^2)}$ in the direction of the vector $k_1\mathbf{i} + k_2\mathbf{j}$. This case can be reduced to the previous one by rotating the x- and y-axes so that the new x-axis is in the direction of the vector $k_1\mathbf{i} + k_2\mathbf{j}$.

(b) $\phi = s/r$, $r^2 = x^2 + y^2 + z^2$, $r > 0$, *s a positive constant.* Laplace's equation is satisfied for $r > 0$, and if we write down the components of $-\nabla\phi$ in spherical polar coordinates [§ 7:8(i)(5)], so that

$$\mathbf{q} = q_r\hat{\mathbf{r}} + q_\theta\hat{\boldsymbol{\theta}} + q_\psi\hat{\boldsymbol{\psi}},$$

we find that

$$q_r = s/r^2, \quad q_\theta = q_\psi = 0.$$

We thus have the case of fluid flowing from the origin in all directions, with speed s/r^2 at distance r from the origin. This corresponds to the case of an isolated point-source at the origin [see § 8:2(iv)]. We note that ϕ is not defined at the origin, and that Laplace's equation and hence the equation of continuity are *not* satisfied there.

The volume T of fluid flowing out through an imaginary sphere S of radius a and centre the origin is $\int_S m\mathbf{q} \,.\, d\mathbf{S}$, which here becomes

$$\int_S \frac{ms}{r^2}\hat{\mathbf{r}} \,.\, d\mathbf{S}.$$

For the sphere S the unit outward normal is $\hat{\mathbf{r}}$, so that $d\mathbf{S} = dS\hat{\mathbf{r}}$. Thus

$$T = \int_S \frac{ms}{r^2}\hat{\mathbf{r}} \,.\, d\mathbf{S} = \int_S \frac{ms}{r^2} dS = \int_S \frac{ms}{a^2} dS$$

since on the surface $r = a$. So

$$T = \frac{ms}{a^2}\int_S dS = \frac{ms}{a^2} \,.\, 4\pi a^2$$

or $$T = 4\pi ms. \tag{23}$$

Comparing with 8:2(15), we see that the velocity potential $\phi = s/r$ corresponds to an *isolated point-source of strength s* at the origin.

Using the Divergence Theorem, we can write T in an alternative form as

$$T = \int_V \operatorname{div}(m\mathbf{q})\, dv,$$

and we know that div $\mathbf{q}$ is *zero everywhere except* at the origin, and so from (23) it follows that at the origin div $\mathbf{q} = 4\pi s$; thus at the source, div $\mathbf{q}$ measures the rate at which fluid is being produced.

Changing the sign of s, the potential $\phi = -s/r$, $s > 0$, gives a sink of strength s at the origin.

(c) $\phi = -s \log \rho$, $\rho^2 = x^2 + y^2$, $\rho > 0$, *s a positive constant.* Again, Laplace's equation is satisfied whenever $\rho > 0$. Using cylindrical polar coordinates ρ, ψ, z, so that

$$\mathbf{q} = q_\rho \hat{\boldsymbol{\rho}} + q_\psi \hat{\boldsymbol{\psi}} + q_z \hat{\mathbf{z}},$$

we find here [from § 7:8 (ii) (14)] that

$$q_\rho = s/\rho, \quad q_\psi = q_z = 0.$$

We now have the case of two-dimensional flow, as defined in § 8:2 (iv), with fluid flowing radially in the x–y plane from the origin, with speed s/ρ at distance ρ from the origin. For motion in the x–y plane only this corresponds to an isolated source at the origin; if we are considering a volume of fluid whose velocity at any point depends only on the coordinates x and y, then this corresponds to a uniform line-source along the z-axis.

For motion in the x–y plane, the fluid area A crossing a circle C of radius a and centre the origin in unit time is given by

$$A = \oint_C \sigma q_\rho \, ds,$$

ds being the element of length of C and hence equal to $\rho\, d\psi$. So

$$A = \int_C \sigma \frac{s}{\rho} \rho \, d\psi = 2\pi s \sigma. \tag{24}$$

Thus we have a *two-dimensional point-source of strength* s at the origin corresponding to the potential $-s\log\rho$.

If we change the sign of s, we have a sink at the origin.

(vii) *A simple example of rotational motion*

We defined rotational motion in § 8:2 (v) as motion in which curl $\mathbf{q} \neq \mathbf{0}$. In such cases, curl $\mathbf{q}$ is the vorticity vector, and we denote it by $\boldsymbol{\zeta}$. We shall consider here a simple example of this type of motion in which a liquid rotates with constant angular velocity ω about a fixed vertical axis Oz; the motion is like that of a solid cylinder rotating about its axis.

If P is the point (x, y, z), in Fig. 50 then the velocity of a liquid

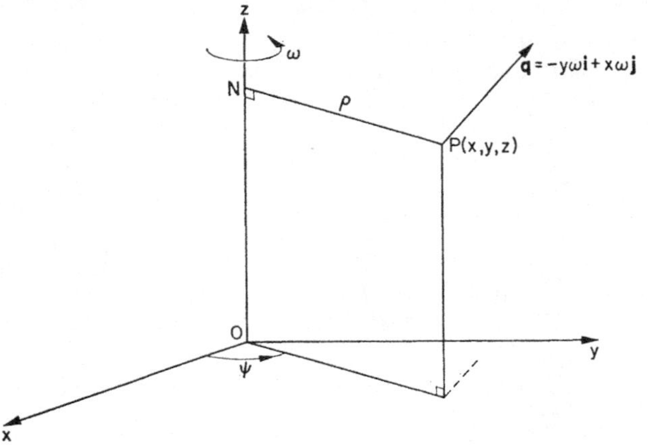

FIG. 50

particle at P is parallel to the x–y plane and perpendicular to NP, and of magnitude ωNP. So

$$\mathbf{q} = \omega(-NP\sin\psi\mathbf{i} + NP\cos\psi\mathbf{j}) = -\omega y\mathbf{i} + \omega x\mathbf{j}, \qquad (25)$$

and the motion is two-dimensional according to the definition in § 8:2 (iv).

For a liquid, the density m is constant, and the equation of continuity requires div $\mathbf{q} = 0$, which is clearly satisfied by the value of $\mathbf{q}$ in (25) above.

The vorticity ζ is given by

$$\zeta = \text{curl}\,\mathbf{q} = \begin{vmatrix} \mathbf{i} & \mathbf{j} & \mathbf{k} \\ \dfrac{\partial}{\partial x} & \dfrac{\partial}{\partial y} & \dfrac{\partial}{\partial z} \\ -\omega y & \omega x & 0 \end{vmatrix} = 2\omega\mathbf{k}.$$

This agrees with the statement in § 8:2 (v) that the angular velocity of a fluid is $\frac{1}{2}$ curl $\mathbf{q}$. In this motion the liquid particles are moving in concentric horizontal circles about the z-axis. The motion is steady, since $\mathbf{q}$ is independent of t.

We could have used cylindrical polar coordinates ρ, ψ, z here, instead of cartesians. Then we should have

$$\mathbf{q} = \omega\rho\hat{\boldsymbol{\psi}},$$

and from 7:8 (15) and (16),

$$\text{div}\,\mathbf{q} = \frac{1}{\rho}\frac{\partial}{\partial\psi}(\omega\rho) = 0,$$

$$\begin{aligned} \text{curl}\,\mathbf{q} &= -\frac{1}{\rho}\frac{\partial}{\partial z}(\omega\rho^2)\hat{\boldsymbol{\rho}} + \frac{1}{\rho}\frac{\partial}{\partial\rho}(\omega\rho^2)\hat{\mathbf{z}} \\ &= 2\omega\hat{\mathbf{z}} = 2\omega\mathbf{k}, \end{aligned}$$

as before.

Let us now calculate the *circulation* in a circular path C of radius a surrounding the origin. Using the definition of § 6:4, with $\mathbf{V} = m\mathbf{q}$, the circulation in C is $\oint_C m\mathbf{q}\,.\,d\mathbf{r}$. We can give a physical interpretation to this line integral if we imagine a pipe of very small cross-section in the shape of the circle C placed in the liquid. Let ds be the element of length on C, and q_s the component of $\mathbf{q}$ in the direction of ds. Then $mq_s ds$ measures the volume of liquid flowing through the element d*s* in unit time. But (see Fig. 51),

$$mq_s \mathrm{d}s = m\mathbf{q} \cdot \mathrm{d}\mathbf{r}.$$

So if we integrate round C, then $\oint_C m\mathbf{q} \cdot \mathrm{d}\mathbf{r}$ measures the volume

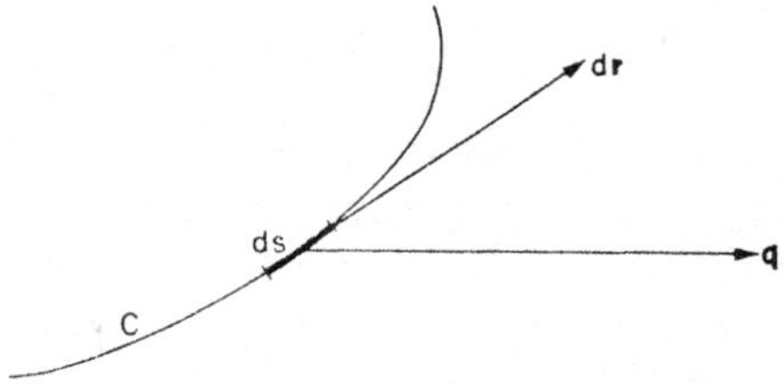

FIG. 51

of liquid flowing round C in unit time; this is the circulation of $m\mathbf{q}$ in the circuit C. Here we have

$$\oint_C m\mathbf{q} \cdot \mathrm{d}\mathbf{r} = \oint_C [m\omega(-y\mathbf{i} + x\mathbf{j}) \cdot (\mathrm{d}x\mathbf{i} + \mathrm{d}y\mathbf{j})]$$

$$= m\omega \oint_C (-y\,\mathrm{d}x + x\,\mathrm{d}y)$$

$$= 2\pi a^2 m\omega, \quad \text{by Green's Theorem [4:6(6)].}$$

In general, for any closed curve within a moving fluid, $\oint_C m\mathbf{q} \cdot \mathrm{d}\mathbf{r}$ measures the volume of fluid flowing round C in unit time.

(viii) *Stokes's Theorem*

If we take any simple closed curve C within a volume of moving fluid, and span it with any open surface S, *lying entirely within the fluid*, then by Stokes's Theorem [§ 7:4 (i)],

$$\oint_C \mathbf{q} \cdot \mathrm{d}\mathbf{r} = \int_S \operatorname{curl} \mathbf{q} \cdot \mathrm{d}\mathbf{S}.$$

It follows at once that if curl $\mathbf{q} = \mathbf{0}$ at every point of S, then $\oint_C \mathbf{q} \cdot d\mathbf{r} = 0$; that is, the circulation of $\mathbf{q}$ in the circuit C is zero. This is true for irrotational motion, for *any* circuit C, *whenever* we can span C by a surface S lying within the fluid. Remembering that for irrotational motion $\mathbf{q} = \nabla\phi$, we thus have

$$\oint_C \mathbf{q} \cdot d\mathbf{r} = \oint_C \nabla\phi \cdot d\mathbf{r} = 0. \tag{26}$$

In § 6:5 we *defined* an irrotational vector field as one in which the circulation is zero in every circuit in the field; the reader will recall that in Chapters 6 and 7 we were concerned only with single-valued scalar and vector point functions, so that ϕ of (26) above is assumed to be single-valued at each point of the field. However, in hydrodynamics, it is necessary to consider cases where ϕ is not single-valued, and in such cases we cannot say that if the motion is irrotational then the circulation in every circuit is zero. This is best illustrated by a particular example; suppose we have fluid moving steadily round the outside of a fixed circular cylinder of radius a and axis Oz, which is vertical, in such a way that the fluid particles are traversing horizontal circles with centres on Oz, the speed at a distance ρ from Oz being a/ρ. We can use cylindrical polar coordinates ρ, ψ, z; the velocity $\mathbf{q}$ at a point $P(\rho, \psi, z)$ is given by $\mathbf{q} = (a/\rho)\hat{\boldsymbol{\psi}}$, and we find easily that curl $\mathbf{q} = \mathbf{0}$, so that the motion is irrotational, in the sense defined in § 8:2 (v).

If we take a curve C in the fluid, such that C does not encircle the origin (see Fig. 52), then we can span C by an open surface S lying in the fluid, and the circulation in C is thus zero, by Stokes's Theorem. But if we take a circle C_1, of radius ρ_1, and centre the origin, as shown in the figure, then it is impossible to span C_1 by an open surface lying entirely in the fluid, and Stokes's Theorem does not apply. We can, however, easily calculate the circulation of $\mathbf{q}$ in C_1; it is

$$\oint_{C_1} \mathbf{q} \cdot d\mathbf{r} = \oint_{C_1} \frac{a}{\rho_1} \hat{\boldsymbol{\psi}} \cdot \rho_1 \, d\psi \hat{\boldsymbol{\psi}}$$

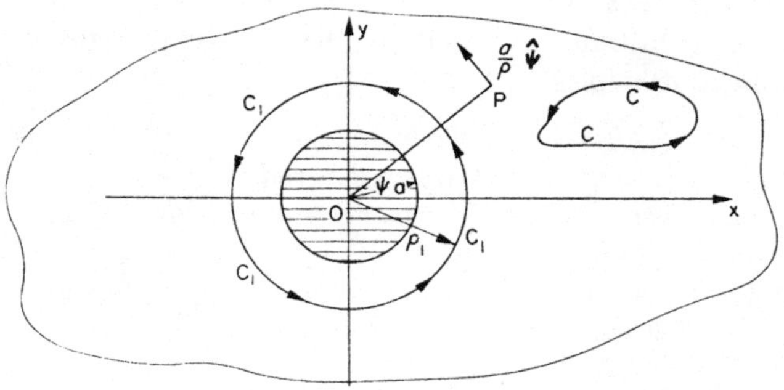

FIG. 52

since in this example the vector **dr** is given by $\rho_1 \mathrm{d}\psi \hat{\boldsymbol{\psi}}$. Hence

$$\oint_{C_1} \mathbf{q} \cdot \mathbf{dr} = \oint_{C_1} a \, \mathrm{d}\psi = 2\pi a.$$

So the circulation in *any* circle centre the origin is the same, and *non-zero*. In this case, ϕ is in fact not single-valued. For

$$\oint_{C_1} \mathbf{q} \cdot \mathbf{dr} = \oint_{C_1} \nabla\phi \cdot \mathbf{dr}$$

$$= \oint_{C_1} \mathrm{d}\phi, \quad \text{by 6:3(4)},$$

$$= [\phi]_{C_1}.$$

Thus the circulation of **q** in C_1 gives us the *change* in ϕ when C_1 is traversed completely. If ϕ were single-valued then this change would be zero. In the present case we know that it is $2\pi a$; ϕ increases by $2\pi a$ every time we traverse the circuit C_1. We note that though ϕ is not single-valued, $\mathbf{q} = \nabla\phi$ is, so that there is no ambiguity in the velocity at any point. So there can exist irrotational motion in which some circuits (at least) have non-zero circulation.

(ix) *Some worked examples*

We now illustrate the results of the preceding sections by a few worked examples.

EXAMPLE 1. *Show that if the motion of an incompressible fluid is such that the vorticity is everywhere the gradient of a scalar function of position, then the velocity components are solutions of Laplace's equation.*

Let $\mathbf{q} = u\mathbf{i} + v\mathbf{j} + w\mathbf{k}$. We are given that $\boldsymbol{\zeta} = \text{curl } \mathbf{q} = \nabla\psi$, where ψ is a scalar point function. So

$$w_y - v_z = \psi_x, \quad u_z - w_x = \psi_y, \quad v_x - u_y = \psi_z, \tag{1}$$

the suffixes denoting partial differentiation.

Since the fluid is incompressible, div $\mathbf{q} = 0$. Hence

$$u_x + v_y + w_z = 0. \tag{2}$$

From (2),
$$u_{xx} + v_{xy} + w_{xz} = 0. \tag{3}$$

Using (1), $\quad v_{yx} = u_{yy} + \psi_{yz}, \quad w_{zx} = u_{zz} - \psi_{zy}.$

Substituting for v_{xy} and w_{xz} in (3), we now get

$$u_{xx} + u_{yy} + \psi_{yz} + u_{zz} - \psi_{zy} = 0,$$

or
$$\nabla^2 u = 0.$$

Similarly, $\nabla^2 v = \nabla^2 w = 0$, as required.

EXAMPLE 2. *A non-viscous incompressible fluid is in motion under the action of an external force* $\mathbf{F}$ *per unit mass. If* $\mathbf{F} = -\text{grad } V$, *show that the vorticity* $\boldsymbol{\zeta}$ *and the particle velocity* $\mathbf{q}$ *are related by the equation*

$$\frac{d\boldsymbol{\zeta}}{dt} = (\boldsymbol{\zeta} \cdot \nabla)\mathbf{q}.$$

The equation of motion under an external force $\mathbf{F}$ is, by § 8:2 (v) (21),

$$\frac{\partial \mathbf{q}}{\partial t} + (\mathbf{q} \cdot \nabla)\mathbf{q} = \mathbf{F} - \frac{1}{m}\nabla p. \tag{4}$$

For an incompressible fluid m is constant and we are given that $\mathbf{F}$ is conservative, so that we can rewrite (4) as

$$\frac{\partial \mathbf{q}}{\partial t} + (\mathbf{q} \cdot \nabla)\mathbf{q} = -\nabla\left(\frac{p}{m} + V\right).$$

Expanding $(\mathbf{q} \cdot \nabla)\mathbf{q}$, we get

$$\frac{\partial \mathbf{q}}{\partial t} + \tfrac{1}{2}\nabla \mathbf{q}^2 - \mathbf{q} \times \boldsymbol{\zeta} = -\nabla\left(\frac{p}{m} + V\right),$$

or

$$\frac{\partial \mathbf{q}}{\partial t} = \mathbf{q} \times \boldsymbol{\zeta} - \nabla\left(\frac{p}{m} + V + \tfrac{1}{2}\mathbf{q}^2\right). \tag{5}$$

Now, curl grad $\mathbf{A} = \mathbf{0}$ for any $\mathbf{A}$; so taking the curl of both sides of (5),

$$\operatorname{curl}\frac{\partial \mathbf{q}}{\partial t} = \operatorname{curl}(\mathbf{q} \times \boldsymbol{\zeta}).$$

We now expand the curl on the right, using the result given by 7:5(14) (worked Example 1), and note that

$$\operatorname{curl}\frac{\partial \mathbf{q}}{\partial t} = \frac{\partial}{\partial t}\operatorname{curl}\mathbf{q} = \frac{\partial \boldsymbol{\zeta}}{\partial t}.$$

Then $\dfrac{\partial \boldsymbol{\zeta}}{\partial t} = (\nabla \cdot \boldsymbol{\zeta})\mathbf{q} - (\nabla \cdot \mathbf{q})\boldsymbol{\zeta} + (\boldsymbol{\zeta} \cdot \nabla)\mathbf{q} - (\mathbf{q} \cdot \nabla)\boldsymbol{\zeta}.$

Since the fluid is incompressible, $\nabla \cdot \mathbf{q} = 0$; since $\boldsymbol{\zeta} = \operatorname{curl}\mathbf{q}$, $\nabla \cdot \boldsymbol{\zeta} = 0$. So

$$\frac{\partial \boldsymbol{\zeta}}{\partial t} = (\boldsymbol{\zeta} \cdot \nabla)\mathbf{q} - (\mathbf{q} \cdot \nabla)\boldsymbol{\zeta}.$$

Hence

$$\frac{\mathrm{d}\boldsymbol{\zeta}}{\mathrm{d}t} \equiv \frac{\partial \boldsymbol{\zeta}}{\partial t} + (\mathbf{q} \cdot \nabla)\boldsymbol{\zeta} = (\boldsymbol{\zeta} \cdot \nabla)\mathbf{q}.$$

EXAMPLE 3. *An incompressible non-viscous fluid is in motion under external forces; if the flow is steady and two-dimensional, and if the external forces are conservative, show that*

$$v\nabla^2 u - u\nabla^2 v = 0,$$

where $\mathbf{q} = u\mathbf{i} + v\mathbf{j}$.

Assuming the external force to be $\mathbf{F} = -\nabla V$, the equation of motion is

$$\frac{\partial \mathbf{q}}{\partial t} + (\mathbf{q}\,.\,\nabla)\mathbf{q} = -\nabla\left(V + \frac{p}{m}\right). \tag{6}$$

Since the flow is steady, $\partial\mathbf{q}/\partial t = \mathbf{0}$; so (6) becomes

$$\mathbf{q} \times \boldsymbol{\zeta} = -\nabla\left(V + \frac{p}{m} + \tfrac{1}{2}q^2\right).$$

Hence $$\operatorname{curl}(\mathbf{q} \times \boldsymbol{\zeta}) = \mathbf{0}.$$

We are given $\mathbf{q} = u\mathbf{i} + v\mathbf{j}$, and we know that div $\mathbf{q} = 0$ for an incompressible fluid. This implies that $u_x = -v_y$.

Now

$$\mathbf{q} \times \boldsymbol{\zeta} = \begin{vmatrix} \mathbf{i} & \mathbf{j} & \mathbf{k} \\ u & v & 0 \\ 0 & 0 & v_x - u_y \end{vmatrix} = v(v_x - u_y)\mathbf{i} - u(v_x - u_y)\mathbf{j}.$$

Hence

$$\operatorname{curl}(\mathbf{q} \times \boldsymbol{\zeta}) = \begin{vmatrix} \mathbf{i} & \mathbf{j} & \mathbf{k} \\ \dfrac{\partial}{\partial x} & \dfrac{\partial}{\partial y} & \dfrac{\partial}{\partial z} \\ v(v_x - u_y), & -u(v_x - u_y), & 0 \end{vmatrix}$$

$$= \mathbf{k}\left[-\frac{\partial}{\partial x}(uv_x - uu_y) - \frac{\partial}{\partial y}(vv_x - vu_y)\right]$$

$$= -\mathbf{k}[u_x v_x + uv_{xx} - u_x u_y - uu_{xy} + v_y v_x +$$

$$+ vv_{yx} - v_y u_y - vu_{yy}].$$

Using the relation $u_x = -v_y$, this reduces to

$$\operatorname{curl}(\mathbf{q} \times \boldsymbol{\zeta}) = [u(u_{xy} - v_{xx}) - v(v_{yx} - u_{yy})]\mathbf{k}.$$

But $\operatorname{curl}(\mathbf{q} \times \boldsymbol{\zeta}) = \mathbf{0}$; so since $u_{xy} = -v_{yy}$, $v_{xy} = -u_{xx}$,

$$u(v_{yy} + v_{xx}) - v(u_{yy} + u_{xx}) \equiv u\nabla^2 v - v\nabla^2 u = 0.$$

Exercises 8:2

1. Establish the converse of worked Example 1, namely that for an incompressible fluid, if the velocity components are solutions of Laplace's equation, then the vorticity is everywhere the gradient of a scalar function of position.

2. Show that in the steady motion of an incompressible fluid under the action of conservative external forces,

$$(\mathbf{q} \cdot \nabla)\boldsymbol{\zeta} = (\boldsymbol{\zeta} \cdot \nabla)\mathbf{q}.$$

[Hint: $\operatorname{curl}(\mathbf{q} \times \boldsymbol{\zeta}) = \mathbf{0}$ in this motion.]

3. If the particle velocity $\mathbf{q}$ is a homogeneous function of the coordinates x, y, z, and if O is the origin, show that, for an incompressible fluid moving under conservative forces, whenever the vorticity $\boldsymbol{\zeta}$ at a point P is parallel to OP, the vector $d\boldsymbol{\zeta}/dt$ at P is parallel to $\mathbf{q}$.

4. A liquid is rotating with angular velocity $\boldsymbol{\Omega}$ about a vertical axis, $\boldsymbol{\Omega}$ being a function of the perpendicular distance ρ from the axis of rotation. Show that if the motion is irrotational, $\boldsymbol{\Omega} = \mu\mathbf{k}/\rho^2$, where μ is a constant and $\mathbf{k}$ the unit vector whose direction is vertical.

5. A homogeneous liquid is in steady motion; cylindrical polar coordinates ρ, ψ, z are chosen and the velocity $\mathbf{q}$ at any point has components u, v, w in this coordinate system. Given that w, $\partial u/\partial z$, $\partial v/\partial z$ vanish everywhere, show that the vorticity components ξ, η, ζ are given by

$$\xi = \eta = 0, \qquad \zeta = \frac{\partial v}{\partial \rho} + \frac{v}{\rho} - \frac{1}{\rho}\frac{\partial u}{\partial \psi}.$$

6. Liquid contained within a circular cylinder rotates about the axis of the cylinder which is vertical. Show that the equation of continuity is satisfied by $\mathbf{q} = \boldsymbol{\omega} \times \mathbf{r}$, where $\boldsymbol{\omega}$ is the angular velocity, which is dependent on t only, and $\mathbf{r}$ is the position vector measured from a point on the axis of rotation.

   If the liquid is moving under external forces which have components $(\alpha x + \beta y, \gamma x + \delta y, 0)$ per unit mass, prove that $d\omega/dt = \frac{1}{2}(\gamma - \beta)$. [Hint: use Euler's equation of motion, and form the curl of both sides.]

7. If q is the speed at any point of a liquid in irrotational motion in two dimensions, prove that

$$\left(\frac{\partial q}{\partial x}\right)^2 + \left(\frac{\partial q}{\partial y}\right)^2 = q\nabla^2 q.$$

Show also that if at any instant q takes the same value at every point, then the direction of motion is the same at every point.

8:3 Newtonian gravitational attraction

(i) *The field of force*

If we have two particles of matter, of masses m_1 and m_2, distance r apart, then Newton's law of gravitation tells us that the force of attraction between them is $\gamma m_1 m_2/r^2$, where γ is the "gravitational constant". If m_1 is at P_1 and m_2 at P_2, then the force of attraction exerted *on* m_2 *by* m_1 can be represented by the vector $-\gamma m_1 m_2 \mathbf{r}/r^3$, where $\mathbf{r} = \overrightarrow{P_1P_2}$; for the magnitude of this vector is $\gamma m_1 m_2/r^2$, and its direction is *from* P_2 *to* P_1 (see Fig. 53).

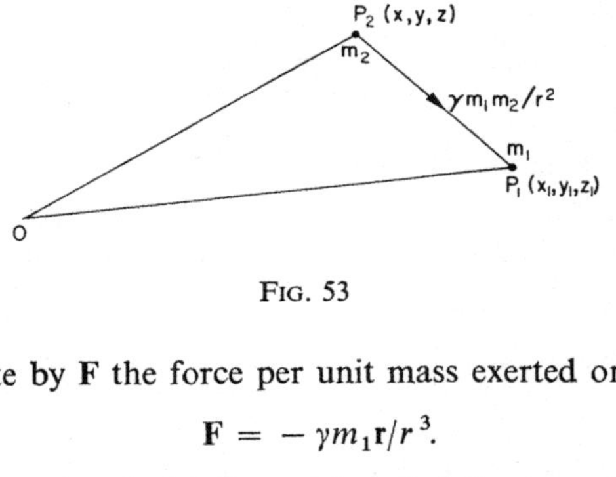

FIG. 53

If we denote by $\mathbf{F}$ the force per unit mass exerted on m_2 by m_1, then

$$\mathbf{F} = -\gamma m_1 \mathbf{r}/r^3. \tag{1}$$

Then $\mathbf{F}$ is associated with P_2 and is called the *intensity of attraction* at P_2. If P_1 is the point (x_1, y_1, z_1) and P_2 the point (x, y, z) referred to fixed cartesian axes, then

$$r^2 = (x - x_1)^2 + (y - y_1)^2 + (z - z_1)^2,$$

$$\mathbf{r} = (x - x_1)\mathbf{i} + (y - y_1)\mathbf{j} + (z - z_1)\mathbf{k}.$$

We can now see that **F** is always irrotational, for all positions of P_2. Using 7:2(3),

$$\nabla \times \mathbf{F} = \nabla \times (-\gamma m_1 r^{-3}\mathbf{r})$$

$$= -\gamma m_1[r^{-3}(\nabla \times \mathbf{r}) + \nabla r^{-3} \times \mathbf{r}];$$

but $\nabla \times \mathbf{r} = \mathbf{0}$ and $\nabla r^{-3} = -3r^{-5}\mathbf{r}$. So

$$\nabla \times \mathbf{F} = -3\gamma m_1 r^{-5}\mathbf{r} \times \mathbf{r} = \mathbf{0}.$$

If we have several particles, then the intensity of attraction at P_2 due to them is obtained simply by taking the vector sum of their individual intensities at P_2; so the resultant intensity of attraction at any point due to a system of particles is irrotational.

Considering a particle of mass m at the origin, we see that the intensity of attraction at any point distant r from the origin is $\mathbf{F} = -\gamma m\mathbf{r}/r^3$. Suppose that we now evaluate $\int \mathbf{F} \cdot \mathrm{d}\mathbf{S}$ over a sphere, centre the origin and radius a. We have

$$\int \mathbf{F} \cdot \mathrm{d}\mathbf{S} = -\gamma m \int \frac{1}{a^3}\mathbf{r} \cdot \hat{\mathbf{r}}\,\mathrm{d}S$$

$$= -\frac{\gamma m}{a^2}\int \hat{\mathbf{r}} \cdot \hat{\mathbf{r}}\,\mathrm{d}S, \quad \text{since } \mathbf{r} = a\hat{\mathbf{r}} \text{ here,}$$

$$= -\frac{\gamma m}{a^2}\int \mathrm{d}S = -4\gamma\pi m.$$

Thus we see that we can compare a particle of mass m at the origin with a sink, or negative source, at the origin in hydrodynamics. Since matter occurs naturally distributed throughout a volume in space or over a surface area, we have to consider the intensity due to volume and surface distributions of attracting material, as well as that due to systems of isolated particles.

It is, in fact, often simpler to deal with continuous distributions by using the gravitational potential, which we shall now define.

(ii) *Gravitational potential*

Since $\mathbf{F}$ is irrotational, there exists a scalar potential function ϕ such that

$$\mathbf{F} = \nabla\phi. \tag{2}$$

For $\mathbf{F}$ given by (1), it is clear that

$$\phi = \gamma m_1/r, \tag{3}$$

since $\nabla(\gamma m_1/r) = -\gamma m_1 \mathbf{r}/r^3$. We say that $\phi = \gamma m_1/r$ is the (*gravitational*) *potential* at the point P_2 due to the particle of mass m_1 at P_1. Since ϕ is a scalar, the potential at P_2 due to a system of particles is obtained by adding together their individual potentials at P_2.

If now we have matter of density m forming a solid body of volume V, we can find the potential at any point P outside the body by means of a volume integral. Let r be the distance of P from an element of volume $\mathrm{d}v$; then

$$\phi = \int_V \frac{\gamma m}{r}\,\mathrm{d}v.$$

For the potential at a point inside the volume this method must be modified, since $1/r$ becomes infinite when $r = 0$; see worked example in § 8:3 (iv).

In a similar way, if matter of surface density σ is distributed over a surface S, then the potential at a point P not on S is given by the surface integral

$$\phi = \int_S \frac{\gamma\sigma}{r}\,\mathrm{d}S.$$

In each case the intensity of attraction at P is given by $\nabla\phi$.

The level surfaces ϕ = constant are the equipotential surfaces. These are level surfaces as defined in § 6:1 since ϕ is single-valued at every point.

Suppose that we have a particle of mass m_1 at the fixed point P and a particle of unit mass at a point P, where $\overrightarrow{P_1P} = \mathbf{r}$. We shall calculate the work done by the attraction $\mathbf{F} = -\gamma m_1\mathbf{r}/r^3$ when the particle at P moves along any path to a new position Q. Let $\overrightarrow{PP'} = d\mathbf{r}$, so that P' is a point close to P on *any* curve PQ joining P to Q (see Fig. 54). The work done by $\mathbf{F}$ in the displacement

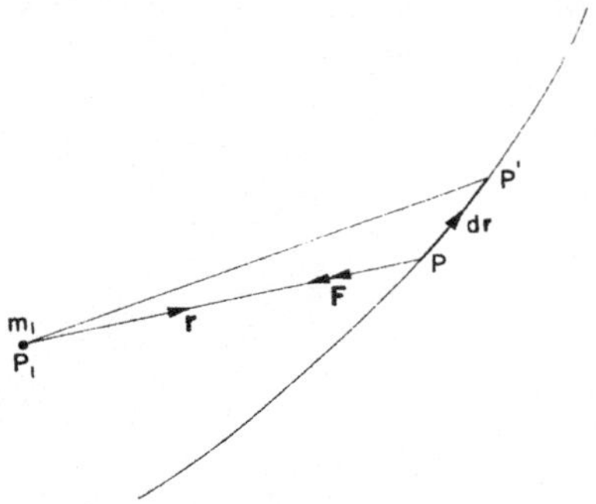

FIG. 54

from P to P' is $\mathbf{F} \,.\, d\mathbf{r}$; hence the work done in the displacement from P to Q is

$$\int_{PQ} \mathbf{F} \,.\, d\mathbf{r} = \int_{PQ} \nabla\phi \,.\, d\mathbf{r}$$

$$= \int_{PQ} d\phi, \quad \text{by 6:3(4)},$$

$$= \phi_Q - \phi_P$$

$$= \frac{\gamma m_1}{P_1Q} - \frac{\gamma m_1}{P_1P},$$

and is the same for all paths, since the line integral is independent of the path of integration. If we now take Q at infinity, so that

$1/P_1Q = 0$, then the work done by the attraction as P is moved to infinity is $-\gamma m_1/P_1P$ or $-\phi_P$. It follows that the potential ϕ at any point P is equal to the work done by the attraction in bringing the particle of unit mass from infinity to the point P.

(iii) *Gauss's Theorem*

We consider first the flux of $\mathbf{F}$ through a simple closed surface S, enclosing a volume V, due to a single particle of mass m_1 at a fixed point P_1, when P_1 lies outside S, as in Fig. 55. Then we

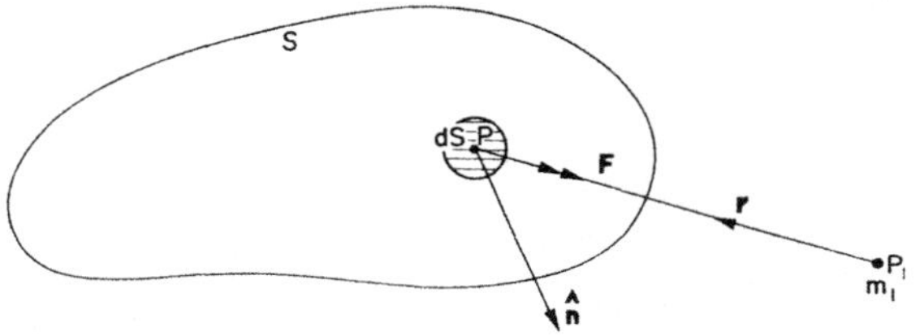

FIG. 55

require $\int_S \mathbf{F} \cdot \mathbf{dS}$, where $\mathbf{F} = -\gamma m_1 \mathbf{r}/r^3$, and $\mathbf{dS} = dS\hat{\mathbf{n}}$, in the usual notation. Using the Divergence Theorem, which applies since $1/r$ is not zero anywhere within V, we have

$$\begin{aligned}\int_S \mathbf{F} \cdot \mathbf{dS} &= \int_V \operatorname{div} \mathbf{F}\, dv \\ &= -\gamma m_1 \int_V [r^{-3}\nabla \cdot \mathbf{r} + \mathbf{r} \cdot \nabla r^{-3}]\, dv \\ &= -\gamma m_1 \int_V [3r^{-3} + \mathbf{r} \cdot 3r^{-5}\mathbf{r}]\, dv \\ &= 0.\end{aligned}$$

By addition, this result can be extended to cover the case of any number of particles lying outside S.

When P_1 lies within S then r^{-3} becomes infinite at the point P_1 and we cannot employ the Divergence Theorem throughout V. To overcome this difficulty we surround P_1 with a small sphere S', centre P_1 and radius a, say, enclosing a volume V'. Then the Divergence Theorem applies to the volume $V-V'$, bounded by the surfaces S and S', and so

$$\int_{S+S'} \mathbf{F} \cdot d\mathbf{S} = 0,$$

the unit normals $\hat{\mathbf{n}}$ being directed away from the volume $V-V'$,

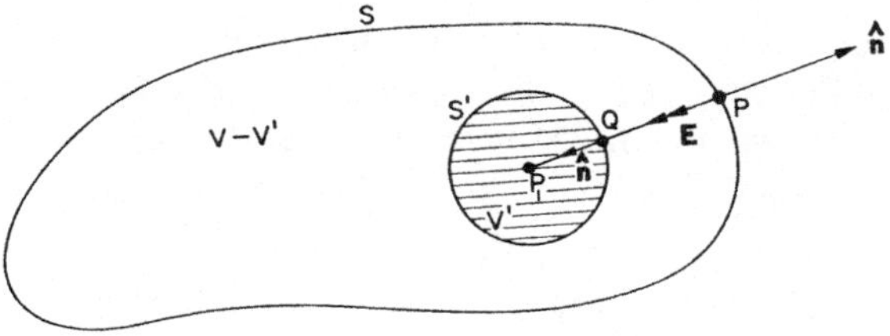

FIG. 56

as shown in Fig. 56. The value of $\int_{S'} \mathbf{F} \cdot d\mathbf{S}$ is given by

$$\int_{S'} \mathbf{F} \cdot d\mathbf{S} = \int_{S'} \frac{\gamma m_1}{a^3} a\hat{\mathbf{r}} \cdot \hat{\mathbf{r}}\, dS,$$

since $\hat{\mathbf{n}}\,(= \hat{\mathbf{r}})$ is in the same direction as $\mathbf{F}$ at any point Q on S'. So

$$\int_{S'} \mathbf{F} \cdot d\mathbf{S} = \frac{\gamma m_1}{a^2} \int_{S'} dS$$

$$= 4\gamma\pi m_1. \tag{4}$$

This is true for all a, however small, and so we can let $a \to 0$. Hence, when $a \to 0$,

$$\int_{S} \mathbf{F} \cdot d\mathbf{S} = \int_{S+S'} \mathbf{F} \cdot d\mathbf{S} - \int_{S'} \mathbf{F} \cdot d\mathbf{S}$$

$$= -4\pi\gamma m_1. \quad \text{[Compare Misc. Ex. 7, No. 18.]}$$

Thus for a particle of mass m_1 within S, the flux of $\mathbf{F}$ through the surface gives $-4\pi\gamma$ times the mass within the surface. This extends at once to any number of particles, by addition, and so we have *Gauss's Theorem*; the flux of the force of attraction through a surface is equal to $-4\pi\gamma$ times the mass enclosed by the surface.

Since $\int_S \mathbf{F} \cdot \mathrm{d}\mathbf{S} = 0$ when there is no matter inside S, it follows from the relation $\mathbf{F} = \nabla\phi$ that

$$\int_S \nabla\phi \cdot \mathrm{d}\mathbf{S} = \int_V \nabla^2\phi \, \mathrm{d}v = 0.$$

This is true for all surfaces S which contain no matter, and so $\nabla^2\phi = 0$ in empty space; that is, *Laplace's equation* is satisfied by the gravitational potential in empty space. It follows from (4) that

$$\int_{S'} \mathbf{F} \cdot \mathrm{d}\mathbf{S} = -4\gamma\pi m_1,$$

the direction of the unit normal $\hat{\mathbf{n}}$ being now outward from S'; this is true for all spheres S' surrounding the point P_1. Hence, letting the radius of the sphere tend to zero, we see that at P_1

$$\nabla^2\phi = -4\pi\gamma m_1.$$

This is true at every point where there is matter; at such points the gravitational potential satisfies *Poisson's equation*

$$\nabla^2\phi = -4\pi\gamma m, \tag{5}$$

where m is the density of the matter at the point. (We can think of Laplace's equation as a special case of Poisson's equation, given by $m = 0$.)

(iv) *Worked examples*

EXAMPLE 1. *Discuss the continuity of the gravitational potential and the intensity across a surface distribution of matter.*

Suppose that we have a distribution of surface density σ on a surface S. Take a point P on the surface, and surround it by a small circle C, of area A and radius δ as shown in Fig. 57. The

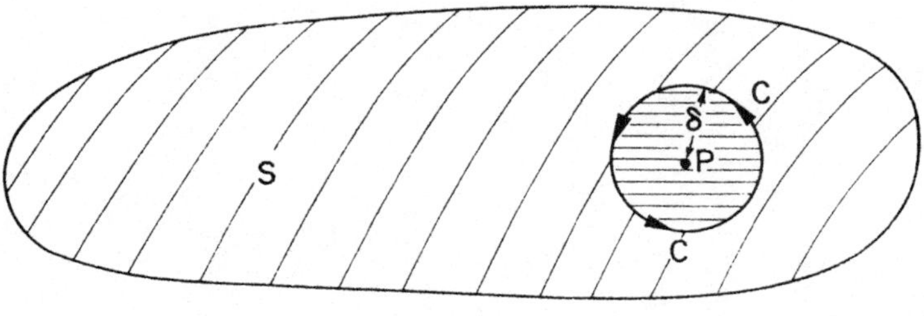

FIG. 57

potential at P due to the matter *on* S and *outside* C can be calculated from the surface integral $\int(\sigma\gamma/r)\mathrm{d}S$ taken over $S - A$. Denote this potential by ϕ; then we *define* the potential ϕ_P at P due to the surface distribution by

$$\phi_P = \lim_{A \to 0} \int_{S-A} \frac{\sigma\gamma}{r}\, \mathrm{d}S.$$

As $A \to 0$, the integrand in $\int(\sigma\gamma/r)\mathrm{d}S$ becomes infinite like $1/\delta$, but $\mathrm{d}S$ is of the order δ^2, and so the integral remains finite.

Further, if $\phi_{P'}$ is the potential at a point P' *not on* S, but *very close* to P, then

$$\phi_{P'} - \phi_P \to 0 \quad \text{as } P' \to P.$$

So the potential function ϕ is continuous across a surface distribution of matter.

Now consider the intensity; enclose a *small* area of the surface S surrounding P by a circular disc, of cross-section $ABCD$, as shown in Fig. 58, the plane faces of the disc being perpendicular to $\hat{\mathbf{n}}$, and the lengths BC, AD *very small* compared with AB, DC. The intensity due to the material of S not enclosed by the disc is clearly continuous as we cross from one side of the surface S to the other at P; for as far as this material is concerned, P is a point in *empty* space.

We apply Gauss's Theorem to the surface of the disc; the contribution of the surface integral over the *curved* surface of the disc is negligible compared with the contributions from the plane

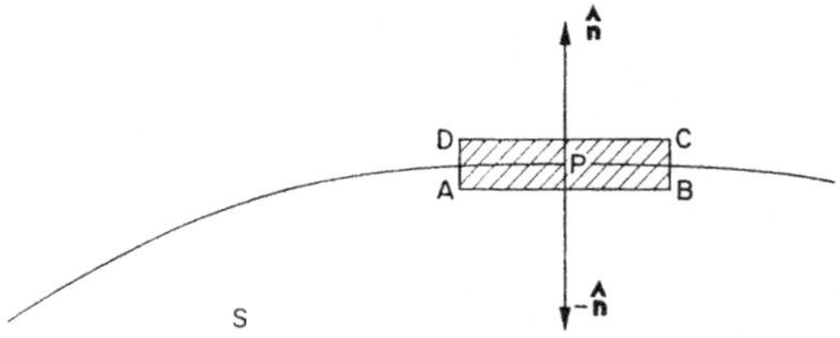

FIG. 58

faces, since we have taken AD, BC very small compared with BA, DC. If $\mathbf{F}$, $\mathbf{F}_1$ are the intensities on the faces CD, AB, then Gauss's Theorem [§ 8:3 (iii)] gives

$$\int \mathbf{F} \,.\, \hat{\mathbf{n}}\, dS + \int \mathbf{F}_1 \,.\, -\hat{\mathbf{n}}\, dS = -4\pi\gamma \int \sigma\, dS. \qquad (1)$$

If we take the circular disc small enough, so that the plane faces are elements of area, then (1) reduces to

$$\mathbf{F} \,.\, \hat{\mathbf{n}} - \mathbf{F}_1 \,.\, \hat{\mathbf{n}} = -4\pi\gamma\sigma. \qquad (2)$$

But $\mathbf{F} = \nabla\phi$; so $\mathbf{F} \,.\, \hat{\mathbf{n}} = \nabla\phi \,.\, \hat{\mathbf{n}} = \partial\phi/\partial n$. Thus if ϕ, ϕ_1 are the potentials on opposite sides of S, (2) becomes

$$\frac{\partial\phi}{\partial n} - \frac{\partial\phi_1}{\partial n} = -4\pi\gamma\sigma.$$

Thus the *normal component* $\mathbf{F} \,.\, \hat{\mathbf{n}}$ of the intensity is *discontinuous* when we cross a surface distribution of matter.

EXAMPLE 2. *Find the potential and intensity of attraction at any point* (*a*) *outside*, (*b*) *within, a thin uniform spherical shell of mass M and radius a.*

Take the origin of coordinates O at the centre of the shell. The matter of the shell is symmetrically placed about O; hence the potential will be a function of r only, where r is the distance from the origin. Let $\mathbf{F}$ be the intensity of attraction and ϕ the potential at any point P due to the shell. Since $\mathbf{F} = \nabla\phi$, we have for any surface,

$$\mathbf{F} \cdot \mathrm{d}\mathbf{S} = \nabla\phi \cdot \hat{\mathbf{n}}\,\mathrm{d}S = \frac{\partial\phi}{\partial n}\mathrm{d}S, \quad \text{by 6:2(10)},$$

where $\partial/\partial n$ denotes differentiation along the outward normal. Applying Gauss's Theorem to a sphere S' with centre O and

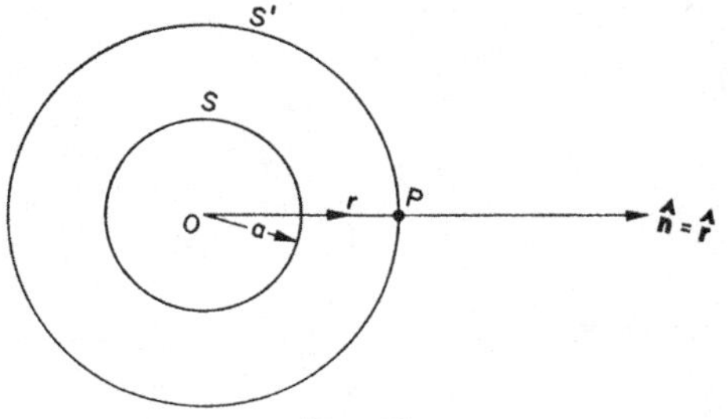

FIG. 59

radius r, as in Fig. 59, so that $\hat{\mathbf{n}} = \hat{\mathbf{r}}$, and $\partial\phi/\partial n = \partial\phi/\partial r$, we have

$$\int_{S'} \mathbf{F} \cdot \mathrm{d}\mathbf{S} = \int_{S'} \frac{\partial\phi}{\partial n}\,\mathrm{d}S = \int_{S'} \frac{\partial\phi}{\partial r}\mathrm{d}S = -4\pi\gamma M, \quad r > a,$$

$$= 0, \quad r < a.$$

In the surface integral, $\partial\phi/\partial r$ is constant for all positions of P on the sphere S', by symmetry about O. So

$$\int_{S'} \frac{\partial\phi}{\partial r}\mathrm{d}S = 4\pi r^2 \frac{\partial\phi}{\partial r} = -4\pi\gamma M, \quad r > a,$$

$$= 0, \quad r < a.$$

Therefore $$\frac{\partial \phi}{\partial r} = -M/r^2, \quad r > a,$$

$$= 0, \quad r < a.$$

Thus $$\phi = M/r + c_1, \quad r > a,$$

$$= c_2, \quad r < a,$$

where c_1, c_2 are constants. To find c_1, we note from § 8:3 (ii) (3) that as $r \to \infty$, $\phi \to 0$; so $c_1 = 0$. Further, ϕ is continuous (see Example 1), taking the value M/a on the sphere S, and hence $c_2 = M/a$.

To find **F**, the intensity, we have $\mathbf{F} = \nabla\phi$. Since ϕ is a function of r only, then by § 7:8 (i) (5),

$$\nabla\phi = \frac{\partial \phi}{\partial r}\hat{\mathbf{r}},$$

and hence $$\mathbf{F} = -\frac{M}{r^2}\hat{\mathbf{r}} = -\frac{M}{r^3}\mathbf{r}, \quad r > a,$$

$$= 0, \quad r < a.$$

We note that the field of force **F** for points outside S is the same as if the whole mass M were concentrated at O.

EXAMPLE 3. *Find the potential and intensity of attraction due to a sphere of mass M, uniform density m and radius a.*

Again, taking O at the centre of the sphere, the gravitational field is symmetrical about O. The sphere can be thought of as made up of thin spherical shells; it follows from Example 2 by summation over all the shells that the potential and intensity at a point P outside the sphere are $\gamma M/r$, $-\gamma M\mathbf{r}/r^3$ respectively, where $\overrightarrow{OP} = \mathbf{r}$.

When P is inside the sphere, we begin by considering the sphere S with the matter removed between concentric spheres of radii

$r \pm \delta$, where δ is small (see Fig. 60). The intensity at P due to the matter between the spheres with radii $r + \delta$ and a is zero, by

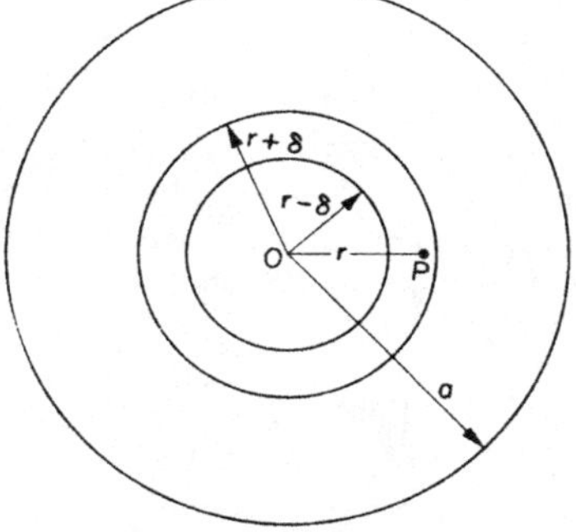

FIG. 60

Example 2. The intensity at P due to the solid sphere of radius $r - \delta$ is given by

$$\mathbf{F} = -m'\gamma\mathbf{r}/r^3,$$

where $$m' = \tfrac{4}{3}\pi m(r-\delta)^3.$$

Hence, at any point P where $\overrightarrow{OP} = \mathbf{r}$,

$$\frac{\partial\phi}{\partial r}\hat{\mathbf{r}} = \mathbf{F} = -\tfrac{4}{3}\pi\gamma m\frac{(r-\delta)^3}{r^3}\mathbf{r}$$

$$= -\tfrac{4}{3}\pi\gamma m\frac{(r-\delta)^3}{r^2}\hat{\mathbf{r}}.$$

Letting $\delta \to 0$, $$\frac{\partial\phi}{\partial r} = -\tfrac{4}{3}\pi\gamma m r.$$

So $$\phi = -\tfrac{2}{3}\pi\gamma m r^2 + c.$$

Since ϕ is continuous, we must have $\phi = M/a$ when $r = a$; thus $c = 2\pi\gamma ma^2$, and

$$\phi = \tfrac{2}{3}\pi\gamma m(3a^2 - r^2), \quad \mathbf{F} = -\tfrac{4}{3}\pi\gamma m\mathbf{r}.$$

Exercises 8:3

1. A body of volume V, bounded by a surface S, is of variable density m, and $\mathbf{r}$ is the position vector of any point within the body or on its surface with respect to the origin O which lies outside the body. By applying the Divergence Theorem to $\int_S \gamma m\hat{\mathbf{r}} \cdot d\mathbf{S}$, show that the potential at O due to the body is given by

$$\frac{2\phi}{\gamma} = \int_S m\hat{\mathbf{r}} \cdot d\mathbf{S} - \int_V (\hat{\mathbf{r}} \cdot \nabla m)\, dv.$$

2. If in the previous example m is constant, and Poisson's equation is satisfied, show that

$$\phi = -\frac{1}{8\pi}\int_S (\nabla \cdot \mathbf{F})\hat{\mathbf{r}} \cdot d\mathbf{S}.$$

[Hint: from Poisson's equation, $m = -\nabla^2\phi/4\pi\gamma$; use the result of 1 above.]

3. A solid sphere, of radius a and centre the origin, is made of material whose density at a distance r from the origin is kr^3. Show that the potential takes the values
   (i) $\gamma M/r$, at any external point,
   (ii) $\gamma M(6a^5 - r^5)/5a^6$, at any internal point,
   r being the distance from the origin and M being the total mass of the sphere. [In (ii), use the method of worked Example 3.]

8:4 Electrostatics

(i) *The field of force and electrostatic potential*

The force on a unit positive charge at a point P, due to a charge e at the origin is called the *electrostatic field*, or the *electric field*, at P. We denote it by the vector $\mathbf{E}$; if $\overrightarrow{OP} = \mathbf{r}$, then, by "Coulomb's law",

$$\mathbf{E} = e\mathbf{r}/r^3.$$

Assuming e to be positive, the electric field at P is thus directed away from the origin. If we have a negative charge $-e$ at the origin, then the direction of the field at P is reversed.

It is clear that $\mathbf{E}$ is irrotational, and we can define the potential V so that

$$\mathbf{E} = -\nabla V.$$

Then the potential at P due to a charge e at the origin is

$$V = e/r.$$

The electrostatic field is very similar to the gravitational one. To find V and $\mathbf{E}$ for a finite number of point charges, we merely add the individual results. We can also have volume and surface distributions of charge, which lead to volume and surface integrals to define V, and hence $\mathbf{E}$.

In electrostatics we have very frequently a pair of equal and opposite charges, close together, forming a dipole (compare the doublet in § 8:2 (iv)). The theory of dipoles is of great importance in physical chemistry, since many molecules are electric dipoles. If we have a pair of charges $+e$, $-e$ at a distance l apart, forming a dipole, then l is small, and e is large compared with l; if $el = m$, then for a dipole, m remains constant (and finite) as $l \to 0$, and we say that m is the *moment* of the dipole. For a dipole placed as in Fig. 61, the potential V at P is

$$\begin{aligned} V &= -\frac{e}{AP} + \frac{e}{BP} \\ &= e\left(\frac{1}{BP} - \frac{1}{AP}\right). \qquad (1) \end{aligned}$$

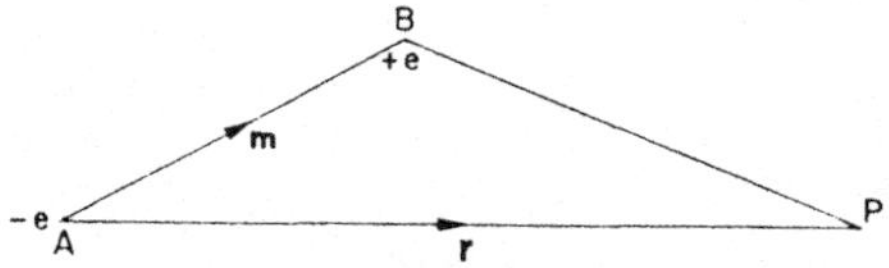

FIG. 61

Denoting AP by r, the quantity $(1/BP - 1/AP)$ can be considered as the change in $1/r$ as we move from A to B. If $\partial/\partial s$ denotes differentiation in the direction AB, then to the first order in l,

$$\frac{1}{BP} - \frac{1}{AP} = l\frac{\partial}{\partial s}\left(\frac{1}{r}\right). \tag{2}$$

We now define the *vector* $\mathbf{m}$ to be in the direction AB, of magnitude el. Then from (2)

$$l\frac{\partial}{\partial s}\left(\frac{1}{r}\right) = l\hat{\mathbf{m}} \cdot \nabla\left(\frac{1}{r}\right), \quad \text{by 6:2(10).} \tag{3}$$

Combining (1), (2) and (3), we can now write

$$V = el\hat{\mathbf{m}} \cdot \nabla\left(\frac{1}{r}\right) = \mathbf{m} \cdot \nabla_A\left(\frac{1}{r}\right), \tag{4}$$

the suffix A implying that the differentiation of $1/r$ is with respect to the coordinates of A. Since.

$$r^2 = (x_A - x_P)^2 + (y_A - y_P)^2 + (z_A - z_P)^2,$$

then simple partial differentiation shows that

$$\nabla_A\left(\frac{1}{r}\right) = -\nabla_P\left(\frac{1}{r}\right).$$

Thus (4) becomes

$$V = -\mathbf{m} \cdot \nabla_P\left(\frac{1}{r}\right) = \frac{\mathbf{m} \cdot \mathbf{r}}{r^3}, \tag{5}$$

where $\overrightarrow{AP} = \mathbf{r}$.

(ii) *Gauss's Law*

We can obtain for the electric field a result very similar to that given by Gauss's Theorem [§ 8:3 (iii)] for the gravitational field, by similar methods. We now have Gauss's Law; for a closed surface S in an electric field **E**,

$$\int_S \mathbf{E} \,.\, d\mathbf{S} = 4\pi \times \text{total charge enclosed by } S.$$

Applying the Divergence Theorem, we find that in space empty of electric charge,

$$\nabla^2 V = 0;$$

that is, the electric potential V satisfies Laplace's equation. If a volume dv has a total charge $m\,dv$, then Poisson's equation is satisfied, so that

$$\nabla^2 V = -4\pi m.$$

Just as in gravitational potential theory, we can show that the normal component of **E** is discontinuous, and changes by $4\pi\sigma$, when we cross a surface distribution of electric charge of density σ.

Exercises 8:4

1. Show that the field **E** at P, where $\overrightarrow{OP} = \mathbf{r}$, due to a dipole of moment **m** at the origin, is given by

$$\mathbf{E} = \frac{3(\mathbf{m} \,.\, \mathbf{r})\mathbf{r}}{r^5} - \frac{\mathbf{m}}{r^3}.$$

2. A dipole of moment **m** is placed at the origin; P is distant r from O, the angle between OP and the direction of **m** is θ. Show that the components of **E** at P in the directions of r and θ increasing are

$$E_r = \frac{2m\cos\theta}{r^3}, \quad E_\theta = \frac{m\sin\theta}{r^3}.$$

3. A point charge e is placed at the point $(0, 0, \frac{5}{12}a)$. Show that the flux of the field **E** due to this charge through the area enclosed by the circle $x^2 + y^2 = a^2$, $z = 0$ is $\frac{16}{13}\pi e$.
[Hint: first write down the potential at any point of the circular area due to the charge. Hence find **E** at any point within the circle; then evaluate $\int \mathbf{E} \,.\, d\mathbf{S}$ over the circle, using polar coordinates.]

8:5 Magnetostatics

We shall mention only the basic results for *small permanent magnets*, which can be treated as *magnetic dipoles*. It is accepted that permanent magnetism, in substances such as iron, is due to the tiny electric currents flowing within individual atoms. It is thought that many atoms are linked together to form a magnetic unit; it is this unit which we can treat as a magnetic dipole. In magnetostatics, only dipoles can occur, since with every positive magnetic pole is associated the equal and opposite (negative) pole.

If we consider a magnetic dipole of moment **m**, then the potential at a point whose position vector with respect to the dipole is **r** is

$$V = \frac{\mathbf{m} \cdot \mathbf{r}}{r^3},$$

just as in the case of an electrostatic dipole [§ 8:4 (i) (5)]. We now define the *magnetic field intensity* **H** by

$$\mathbf{H} = -\nabla V,$$

so that **H** is irrotational. It follows that we can write

$$\mathbf{H} = \frac{3(\mathbf{m} \cdot \mathbf{r})\mathbf{r}}{r^5} - \frac{\mathbf{m}}{r^3}, \tag{1}$$

or, using radial and transverse components,

$$H_r = \frac{2m \cos \theta}{r^3}, \quad H_\theta = \frac{m \sin \theta}{r^3};$$

see Exercises 8:4, Nos. 1 and 2.

In § 8:4 (ii) we saw that for any closed surface S in an electric field **E**,

$$\int_S \mathbf{E} \cdot d\mathbf{S} = 4\pi \times \text{total charge within } S.$$

The corresponding result for the magnetic field **H** is

$$\int_S \mathbf{H} \cdot d\mathbf{S} = 0,$$

since in the magnetic field only dipoles occur, and to every positive magnetic pole corresponds the equal and opposite one.

The reader will notice that the field intensities in Newtonian gravitational theory, electrostatics and magnetostatics are *all* irrotational, and the corresponding potential functions are single-valued.

Exercises 8:5

1. Show that the vector $\mathbf{A} = (\mathbf{m} \times \mathbf{r})/r^3$, where $\mathbf{m}$ is a constant vector and $\mathbf{r}$ the position vector with respect to the origin of the point (x, y, z) satisfies the equations
$$\operatorname{div} \mathbf{A} = 0, \quad \operatorname{curl} \mathbf{A} = \mathbf{H},$$
where **H** is given by 8:5(1).
[**A** is the *magnetic vector potential* corresponding to the magnetic intensity **H**.]

2. The couple on a small magnet of moment **m** due to an external magnetic field $\mathbf{H}'$ is $\mathbf{m} \times \mathbf{H}'$. Deduce that the couple **G** exerted on a magnet of moment **m** by a second magnet of moment $\mathbf{m}'$ distant r from the first is
$$\mathbf{G} = \frac{\mathbf{m} \times \mathbf{m}'}{r^3} - \frac{3(\mathbf{m} \times \mathbf{r})(\mathbf{m}' \cdot \mathbf{r})}{r^5}.$$
Hence show that if two small bar magnets are free to rotate in a plane about their (fixed) centres, then there is a position of equilibrium in which both magnets point in the same direction along their line of centres. (In equilibrium, the couple $\mathbf{G} = \mathbf{0}$.)

8:6 Electromagnetism

(i) *Magnetic field due to a steady electric current*

It was found in 1820 by Oersted that when electric currents flow within two separate conducting systems, these systems exert forces on each other which are similar to those exerted by permanent magnets. Ampère found that two small coils in which currents are flowing exert on each other forces of the same type as those between two electrical dipoles. If a wire carrying a steady current i is in the form of a closed loop, enclosing a small area dS, then we define the magnetic moment $\mathbf{m}$ of the loop by

$$\mathbf{m} = i\mathrm{d}\mathbf{S} = i\mathrm{d}S\hat{\mathbf{n}}, \tag{1}$$

$\mathrm{d}\mathbf{S}$ being the vector area associated with the scalar element $\mathrm{d}S$. This definition implies that $\mathbf{m}$ is in the direction of the normal $\hat{\mathbf{n}}$ to the plane of $\mathrm{d}S$; the accepted convention is that the current i is flowing in the direction of a right-handed rotation about the direction of $\hat{\mathbf{n}}$, as in Fig. 62.

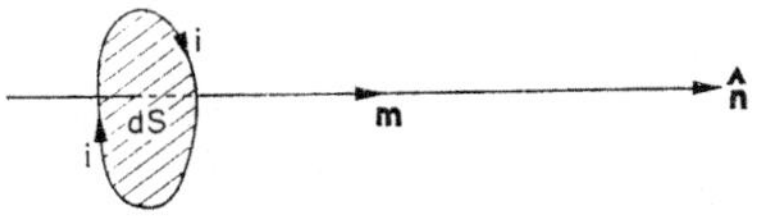

FIG. 62

We can now develop the theory of electromagnetism *by treating every small coil carrying a current i as if it were a dipole of moment* $\mathbf{m}$. It follows at once (see § 8:5) that the magnetic potential at any point P distant r from the centre of a small coil is given by

$$\Omega_P = \frac{\mathbf{m}\,.\,\mathbf{r}}{r^3} = \frac{m\cos\theta}{r^2}, \tag{2}$$

where θ is the angle between the directions of $\mathbf{r}$ and $\mathbf{m}$.

This leads at once to the definition of the magnetic field due to the current in the small coil as

$$\mathbf{H}_P = -\operatorname{grad}\Omega_P. \tag{3}$$

The result in (2) can be extended easily to a large coil; suppose we have a wire in the shape of a closed curve γ, carrying a steady current i. We can span γ by any open surface S, and divide S into a number of small areas by drawing lines on S as shown in Fig. 63. Suppose that $ABCD$ is such an area; if we imagine the same current i to flow around $ABCD$ and every other small area on S in the direction marked by the arrows in Fig. 63, then for lines such

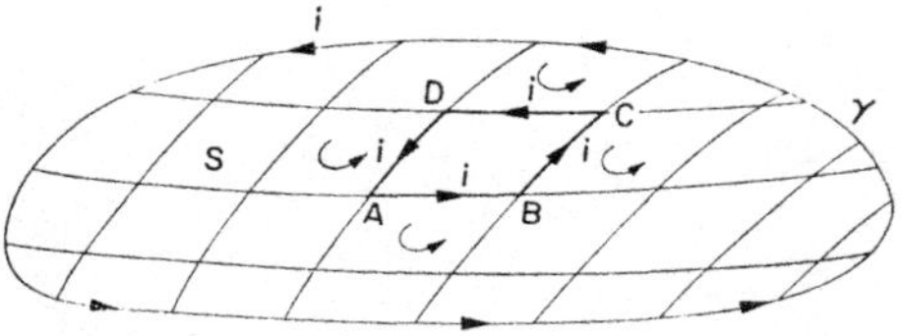

FIG. 63

as AB, the resultant flow will be zero. This is true for all lines except the bounding curve γ of S. So the current i in the large coil γ has the same effect as a current i in each of the small coils. It follows from (2) by integration that the magnetic potential at a point P due to the current in the large coil is

$$\Omega_P = \int_S \frac{i \cos\theta \, dS}{r^2}, \tag{4}$$

since $m = i\mathrm{d}S$.

The integral in (4) is most easily dealt with if we introduce the concept of the *solid angle*. Consider the element of area $\mathrm{d}S$ enclosed by $ABCD$; if we join up every point of $ABCD$ to the point P we shall have a cone, vertex at P, as in Fig. 64. The area $\mathrm{d}\omega$ cut off on a unit sphere, centre P, by the cone is called the *solid angle* $\mathrm{d}\omega$ subtended by the area $ABCD$ at P. Let Q be any point within $ABCD$; and let the angle between PQ and $\mathbf{dS}$ be θ. Then the projection of $\mathrm{d}S$ on a plane perpendicular to PQ at Q is $\mathrm{d}S\cos\theta$; this is the area cut off by the cone on a sphere, centre P and radius $PQ = r$. Hence the area cut off on the unit sphere, centre P is $\mathrm{d}S\cos\theta/r^2$. So

$$\mathrm{d}\omega = \frac{1}{r^2}\mathrm{d}S\cos\theta,$$

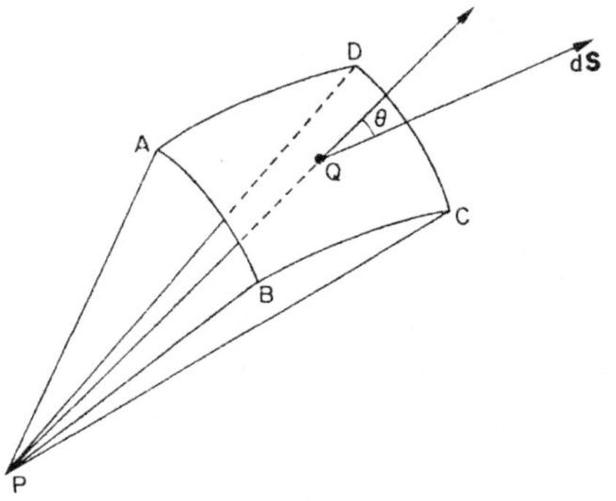

FIG. 64

$d\omega$ being positive or negative according as $\cos\theta$ is positive or negative.

With this definition of $d\omega$, (4) gives

$$\Omega_P = \int_S \frac{i \cos\theta \, dS}{r^2} = \int_S i \, d\omega.$$

Therefore $$\Omega_P = i\omega_P, \tag{5}$$

where ω_P is the solid angle subtended at P by the surface S, and hence by the wire γ.

In (5) ω_P depends on the shape of γ; if we keep γ constant and let P move, then ω_P varies. If we let P trace out some closed path γ', which does not intersect S, then ω_P changes continuously and returns to its original value when P returns to its starting point. If, however, we take a path γ' which intersects S as shown in Fig. 65, then ω_P decreases by 4π as we cross S (see Exercises 8:6, No. 1) and so Ω_P changes by $-4\pi i$. For the directions of γ' and i shown in the diagram, we say that "γ' threads the wire circuit positively". We have now seen that the potential Ω_P is *not* a single-valued function; this is similar to the case of the velocity potential discussed in § 8:2 (viii).

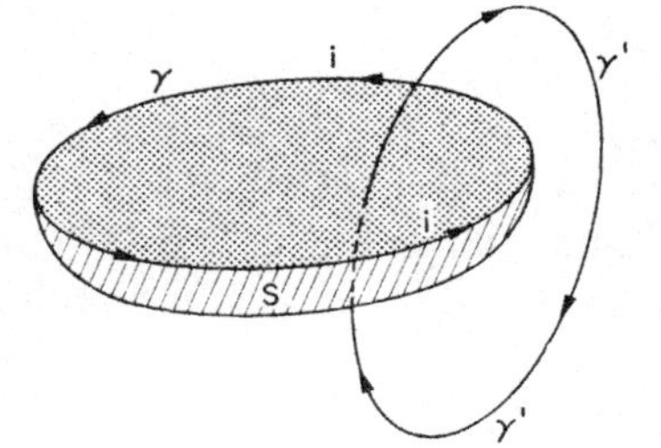

FIG. 65

Using the fact that $\mathbf{H} = -\nabla\Omega$, we can calculate the circulation of $\mathbf{H}$ in any circuit γ'. For

$$\int_{\gamma'} \mathbf{H} \cdot \mathrm{d}\mathbf{r} = \int_{\gamma'} -\nabla\Omega \cdot \mathrm{d}\mathbf{r} = -\int_{\gamma'} \mathrm{d}\Omega$$

$$\left.\begin{aligned} &= -(-4\pi i) = 4\pi i, \quad \text{w en } \gamma' \text{ threads } \gamma, \\ &= 0, \quad \text{for all other circuits } \gamma'. \end{aligned}\right\} \tag{6}$$

If γ' threads γ positively n times, then the circulation of $\mathbf{H}$ in γ' is $4\pi ni$. Again, compare § 8:2 (viii).

(ii) *Field due to a straight steady current*

We saw in § 8:5 that in the magnetic field $\mathbf{H}$

$$\int_S \mathbf{H} \cdot \mathrm{d}\mathbf{S} = 0$$

for any closed surface S. Applying the Divergence Theorem, we obtain

$$\int_V \operatorname{div} \mathbf{H}\, \mathrm{d}v = 0,$$

where V is the volume enclosed by S. This is true for any surface S, and so

$$\operatorname{div} \mathbf{H} = 0. \tag{7}$$

This means that **H** is a *solenoidal vector field*, as defined in § 6:9. A *solenoid* is used to obtain large magnetic fields; it consists essentially of many similar coils lying side by side with their centres on the same straight line. In practice it is formed by winding a wire closely on to a circular cylinder; current then flows through the wire, and a large magnetic field **H** is produced. This field **H** is such that div **H** = 0; this is the origin of the term "solenoidal vector field".

If we take any circuit C which does not thread the circuit in which current is flowing, then

$$\int_C \mathbf{H} \, . \, d\mathbf{r} = 0,$$

and hence, applying Stokes's Theorem to any surface spanning C, but not crossing the current,

$$\int_S \operatorname{curl} \mathbf{H} \, . \, d\mathbf{r} = 0.$$

Since this is true for any surface S spanning C, we have, in space empty of electric current,

$$\operatorname{curl} \mathbf{H} = \mathbf{0}. \tag{8}$$

The relation (8) is, of course, the one we expect from our definition of **H** as $-\nabla\Omega$; it must be stressed that it breaks down in any region occupied by electric currents; in such regions Ω must be replaced by the vector potential **A**, as defined in § 8:6 (iii).

For space empty of electric currents, we have, by (7), since $\mathbf{H} = -\nabla\Omega$,

$$\nabla^2\Omega = 0; \tag{9}$$

that is, Laplace's equation is satisfied by the potential Ω.

By taking particular solutions of (9), we obtain the potential and hence the field **H** corresponding to certain current distributions. We can find the field due to a current flowing in a straight wire if we transform (9) to cylindrical polar coordinates

ρ, ψ, z. Using the result in § 7:8 (ii) (17), (9) above becomes

$$\frac{\partial^2\Omega}{\partial\rho^2}+\frac{1}{\rho}\frac{\partial\Omega}{\partial\rho}+\frac{1}{\rho^2}\frac{\partial^2\Omega}{\partial\psi^2}+\frac{\partial^2\Omega}{\partial z^2}=0.$$

A very simple solution of this equation is $\Omega = A\psi$, where A is a constant. Since $\mathbf{H} = -\nabla\Omega$, then by § 7:8 (ii) (14),

$$H_\rho = 0,\quad H_\psi = -A/\rho,\quad H_z = 0.$$

For a closed circuit C consisting of a circle in any plane parallel to the x–y plane, with its centre on the z-axis and radius ρ_0, we now have

$$\int_C \mathbf{H}\,.\,\mathrm{d}\mathbf{r} = -\int_0^{2\pi}\frac{A}{\rho_0}\rho_0\,\mathrm{d}\psi = -2\pi A$$

$$= 4\pi i,\quad \text{by § 8:6 (i) (6).}$$

Therefore $i = \frac{1}{2}A$.

This is true for *all* values of ρ_0. Further, for any closed circuit C which does not encircle the z-axis, the change in Ω when the circuit is traversed is zero. It follows that $\Omega = A\psi$ corresponds to a current flowing along the z-axis. Thus if a current i flows along an infinitely long wire in the positive direction of the z-axis, we must have

$$H_\rho = 0,\quad H_\psi = 2i/\rho,\quad H_z = 0$$

and $$\Omega = -2i\psi.$$

[The reader should compare the above analysis with that in the example of fluid motion discussed in § 8:2 (viii); the field $\mathbf{H}$ is irrotational, in the sense defined in § 8:2 (v), since curl $\mathbf{H} = \mathbf{0}$, but the circulation of $\mathbf{H}$ is not zero in every circuit in the field. The same holds true for the velocity field $\mathbf{q}$ in the fluid motion example.]

(iii) *The vector potential for a steady current*

So far we have been dealing with the magnetic field $\mathbf{H}$ due to the flow of electric current in isolated wire circuits; now we turn to the case of current flowing through a conducting medium, so that the relation curl $\mathbf{H} = \mathbf{0}$ no longer holds. We can define the *conduction current vector* $\mathbf{j}$ at any point P in a conductor by

$$\mathbf{j} = Ne\mathbf{v},$$

where N is the number of charged particles per unit volume, e is the charge on each and $\mathbf{v}$ is the average velocity of the charges at P. Current then flows in the direction given by the vector $\mathbf{j}$. We consider the case where the current is *steady*, that is, it may vary from point to point in the conductor, but it is independent of time, so that $\mathbf{j}$ is a vector function of x, y, z, but not of t.

Suppose we now take a closed path C in the conductor. Then the circulation of $\mathbf{H}$ in C due to the conduction current is $\int_C \mathbf{H} \cdot d\mathbf{r}$, and it can be shown [C. E. Weatherburn, *Advanced Vector Analysis*, Bell, 1960] that the value of this line integral is 4π times the total current enclosed by C. This total current is equal to the flux of $\mathbf{j}$ through any open surface S spanning C. Thus

$$\oint_C \mathbf{H} \cdot d\mathbf{r} = 4\pi \int_S \mathbf{j} \cdot d\mathbf{S}. \tag{10}$$

Applying Stokes's Theorem to the left-hand side of (10), we find

$$\int_S \operatorname{curl} \mathbf{H} \cdot d\mathbf{S} = 4\pi \int_S \mathbf{j} \cdot d\mathbf{S}.$$

So

$$\int_S (\operatorname{curl} \mathbf{H} - 4\pi\mathbf{j}) \cdot d\mathbf{S} = 0.$$

This is true for *any* surface S spanning C, and so

$$\operatorname{curl} \mathbf{H} = 4\pi\mathbf{j}. \tag{11}$$

Since div curl $\mathbf{V} = 0$ for any vector point function $\mathbf{V}$, it follows that

$$\operatorname{div} \mathbf{j} = 0.$$

[This is analogous to the equation of continuity div $\mathbf{q} = 0$ for an incompressible fluid flow; see § 8:2 (iii).]

The relation div $\mathbf{H} = 0$ still holds for the magnetic field due to the flow of current in a conductor; this is the necessary and sufficient condition (Miscellaneous Exercises 6, No. 4) to enable us to find a vector $\mathbf{A}$ such that

$$\mathbf{H} = \operatorname{curl} \mathbf{A}. \tag{12}$$

The vector $\mathbf{A}$ so defined is not unique, for to it we can add any vector of the form grad ϕ, without altering the value of $\mathbf{H}$ (since curl grad $\phi = \mathbf{0}$). We use the extra condition div $\mathbf{A} = 0$ to make $\mathbf{A}$ unique. The vector $\mathbf{A}$ is the *magnetic vector potential*, or the *vector potential*, corresponding to the current $\mathbf{j}$. [This has already been mentioned in Exercises 8:5, No. 1.]

For any vector $\mathbf{V}$ we have from § 7:4, worked Example 2, the identity

$$\operatorname{curl} \operatorname{curl} \mathbf{V} = \operatorname{grad} \operatorname{div} \mathbf{V} - \nabla^2 \mathbf{V}.$$

Writing $\mathbf{V} = \mathbf{A}$, and using (12) and (11),

$$\operatorname{curl} \operatorname{curl} \mathbf{A} = \operatorname{curl} \mathbf{H} = 4\pi \mathbf{j} = -\nabla^2 \mathbf{A},$$

since div $\mathbf{A} = 0$, by definition. Thus the relation between $\mathbf{A}$ and $\mathbf{j}$ is

$$\nabla^2 \mathbf{A} = -4\pi \mathbf{j}. \tag{13}$$

Equation (13) is a *vector* equation; each *component* of $\mathbf{A}$ obeys an equation similar to Poisson's equation [see § 8:4 (ii)].

We have been discussing only the magnetic effects of *steady* currents. When the current vector, and hence the field vector $\mathbf{H}$, is a function of time t as well as of the spatial coordinates x, y, z, the above equations have to be modified. These more general equations are given in the next section.

(iv) *Maxwell's field equations*

In 1831 Faraday discovered that if a closed wire circuit, *not connected* to any source of electric current, is moved across a magnetic field, then a current flows in the circuit; this current flows *only* when the circuit is *moving*. Also, if the wire circuit is kept stationary and the magnetic field changed, a current again flows in the wire, *as long* as the field is in motion. We say that a current has been *induced* in the wire. Again, if we have a current flowing in one circuit, and we *change* the current flowing in a second nearby circuit, then an additional (induced) current will be detected in the first circuit, as long as the second current is changing.

Clearly, the equations describing these and similar effects will depend on time as well as on the spatial coordinates. Also they will have to take into account both the magnetic field **H** and the electric field **E**. Our equations must relate these two fields; for example, in the case when a closed circuit is moved across a magnetic field **H**, the flow of current in the circuit, being essentially a motion of electric charges, produces an (induced) electric field **E**. The two fields together constitute the *electromagnetic field.*

The basic equations for the electromagnetic field in a charge-free, non-conducting medium are due to Maxwell. We quote them below; we cannot here discuss their derivation, and for a full discussion the reader is referred to C. A. Coulson, *Electricity*, Oliver & Boyd, 1958, and V. C. A. Ferraro, *Electromagnetic Theory*, Athlone Press, 1956.

We suppose the medium to be of constant permeability μ, with dielectric constant K. Then **E** and **H** are both vector functions of x, y, z and t, and Maxwell's equations are

$$\left.\begin{aligned} &\operatorname{div}\mathbf{H} = 0, \quad \operatorname{div}\mathbf{E} = 0, \\ &\operatorname{curl}\mathbf{H} = \frac{K}{c}\frac{\partial \mathbf{E}}{\partial t}, \quad \operatorname{curl}\mathbf{E} = -\frac{\mu}{c}\frac{\partial \mathbf{H}}{\partial t}, \end{aligned}\right\} \tag{14}$$

c being the ratio of the electromagnetic unit of current to the electrostatic unit.

Using equations (14), we can easily show that both **E** and **H** satisfy the standard equations of *wave motion*. Starting with

$$\operatorname{curl}\mathbf{H} = \frac{K}{c}\frac{\partial\mathbf{E}}{\partial t},$$

and taking the curl of both sides,

$$\begin{aligned}\operatorname{curl}\operatorname{curl}\mathbf{H} &= \operatorname{grad}\operatorname{div}\mathbf{H} - \nabla^2\mathbf{H} \quad [\S\ 7{:}4]\\ &= \operatorname{curl}\frac{K}{c}\frac{\partial\mathbf{E}}{\partial t} = \frac{K}{c}\frac{\partial}{\partial t}(\operatorname{curl}\mathbf{E}).\end{aligned}$$

Since $\operatorname{div}\mathbf{H} = 0$ and $\operatorname{curl}\mathbf{E} = -\dfrac{\mu}{c}\dfrac{\partial\mathbf{H}}{\partial t}$, this gives

$$-\nabla^2\mathbf{H} = \frac{K}{c}\frac{\partial}{\partial t}\left[-\frac{\mu}{c}\frac{\partial\mathbf{H}}{\partial t}\right]$$

or

$$\nabla^2\mathbf{H} = \frac{K\mu}{c^2}\frac{\partial^2\mathbf{H}}{\partial t^2}. \qquad (15)$$

Similarly,

$$\nabla^2\mathbf{E} = \frac{K\mu}{c^2}\frac{\partial^2\mathbf{E}}{\partial t^2}. \qquad (16)$$

Equations (15) and (16) show that both **E** and **H** satisfy the wave equation

$$\nabla^2\mathbf{V} = m^2\frac{\partial^2\mathbf{V}}{\partial t^2},$$

m being a constant. They thus show that electromagnetic waves travel with velocity $c/\sqrt{(K\mu)}$; in a vacuum (that is, in space empty of all matter, or "*free space*"), $\mu = K = 1$, and the wave velocity reduces to c. It is found by experiment that in c.g.s. units $c = 2{\cdot}998 \times 10^{10}$; this is also the *velocity of light* in free space, as determined experimentally. Thus Maxwell's equations led to the very important theory that light waves may be electromagnetic in origin; this theory has since been confirmed.

For a conducting medium of conductivity σ, the basic equations (14) have to be modified; this produces a consequential modification in (15) and (16), and the resultant equation for $\mathbf{E}$ is

$$\nabla^2 \mathbf{E} = \frac{K\mu}{c^2}\frac{\partial^2 \mathbf{E}}{\partial t^2} + \frac{4\pi\sigma\mu}{c}\frac{\partial \mathbf{E}}{\partial t}. \tag{17}$$

An exactly similar equation holds for $\mathbf{H}$. Equation (17) is known as the *equation of telegraphy.*

(v) *Some worked examples*

EXAMPLE 1. *Find the field $\mathbf{H}_P$ at a point P on the axis of a solenoid, which has n turns of wire per unit length, the wire carrying a current i. Hence show that the field inside a long solenoid is of magnitude $4\pi ni$.*

Suppose that the wire is wound on a cylinder of radius a; consider one turn of wire; the potential Ω_P at a point P on the axis of the solenoid due to the current in this one loop is, by § 8:6 (i) (5),

$$\Omega_P = i\omega_P,$$

where ω_P is the solid angle subtended by the loop of wire at P.

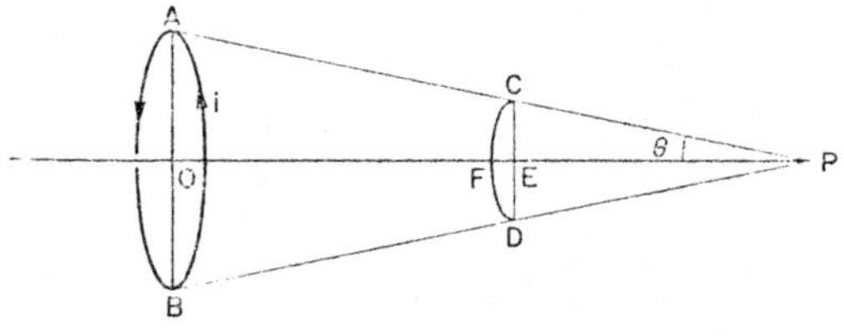

FIG. 66

If the angle subtended by the diameter AOB of the loop at P is 2θ (see Fig. 66), then ω_P is the area of the cap of the unit sphere, centre P, cut off by the cone of semi-vertical angle θ. This area $CFDE$ is shown in section in Fig. 66. Since we know that the surface area cut from a sphere by two parallel planes is equal to the surface cut off by these planes from the circumscribing cylinder which has its generators perpendicular to the planes, we find here

$$\omega_P = 2\pi(1 - \cos\theta).$$

[The radius of the circumscribing cylinder is unity, the planes are the distance FE apart.] Therefore

$$\Omega_P = 2\pi i(1 - \cos\theta) = 2\pi i[1 - x/(a^2 + x^2)^{\frac{1}{2}}],$$

where $x = OP$. Then the field at P due to the single loop of wire is, by symmetry, in the direction OP, and since $\mathbf{H}_P = -\nabla\Omega_P$, its magnitude is

$$H_P = -\frac{\partial\Omega_P}{\partial x} = 2\pi i a^2/(a^2 + x^2)^{\frac{3}{2}}.$$

If θ varies from α at the end of the solenoid nearer to P to β at the other end, we find H_P for the whole solenoid by integration. Thus

$$H_P = \int_{\theta=\beta}^{\theta=\alpha} \frac{2\pi i a^2 n}{(a^2 + x^2)^{\frac{3}{2}}}\,\mathrm{d}x,$$

where $x = a\cot\theta$. Then $\mathrm{d}x = -a\,\mathrm{cosec}^2\theta\,\mathrm{d}\theta$, implying that θ decreases as x increases; so if we take θ increasing from β to α we require

$$H_P = \int_\beta^\alpha \frac{2\pi i a^2 n(+\,a\,\mathrm{cosec}^2\theta)}{a^3\,\mathrm{cosec}^3\theta}\,\mathrm{d}\theta$$

$$= 2\pi n i\int_\beta^\alpha \sin\theta\,\mathrm{d}\theta$$

$$= 2\pi n i(\cos\beta - \cos\alpha).$$

Now suppose the solenoid is very long, and that P is a point on its axis inside it. Then $\beta \to 0$ and $\alpha \to \pi$ as we increase the length of the solenoid indefinitely, and so

$$H_P = 4\pi ni.$$

EXAMPLE 2. *Assuming that the vector potential* **A** *at a point P due to a current i flowing in a closed circuit C is given by*

$$\mathbf{A} = i\oint_C \frac{\mathbf{dr}}{r},$$

where r is the distance from any vector element **dr** *of C to P, show that the field* **H** *at the point P is given by*

$$\mathbf{H} = i\oint_C \frac{\mathbf{dr} \times \mathbf{r}}{r^3}.$$

[For the equation defining **A** here, see the book by Coulson quoted in § 8:6 (iv).]

We know that $\mathbf{H} = \text{curl } \mathbf{A}$, and the differential operators involved in the curl refer to the point P. Therefore

$$\mathbf{H} = \text{curl}\,\mathbf{A} = \text{curl}\left[i\oint_C \frac{\mathbf{dr}}{r}\right] = i\oint_C \text{curl}\left(\frac{\mathbf{dr}}{r}\right);$$

but $$\text{curl}\left(\frac{\mathbf{dr}}{r}\right) = \nabla \times \left(\frac{1}{r}\mathbf{dr}\right) = \frac{1}{r}\nabla \times \mathbf{dr} + \nabla\left(\frac{1}{r}\right) \times \mathbf{dr}.$$

Since the coordinates of P do *not* occur in **dr**, $\nabla \times \mathbf{dr} = \mathbf{0}$.

Further, $$\nabla\left(\frac{1}{r}\right) = -\frac{1}{r^3}\mathbf{r}.$$

So $$\text{curl}\left(\frac{\mathbf{dr}}{r}\right) = -\frac{1}{r^3}\mathbf{r} \times \mathbf{dr} = \frac{1}{r^3}(\mathbf{dr} \times \mathbf{r}).$$

Hence $$\mathbf{H} = i\oint_C \frac{\mathbf{dr} \times \mathbf{r}}{r^3}.$$

EXAMPLE 3. *If ϕ is a scalar function of x, y, z and t, and if* **k** *is the unit vector in the direction of the z-axis, show that*

$$\mathbf{E} = (\mathbf{k}\,.\,\nabla)\nabla\phi - \frac{1}{c^2}\frac{\partial^2\phi}{\partial t^2}\mathbf{k}, \quad \mathbf{H} = \frac{1}{c}\frac{\partial}{\partial t}(\nabla\phi \times \mathbf{k})$$

satisfy Maxwell's equations in free space provided that

$$\frac{\partial^2\phi}{\partial t^2} = c^2\nabla^2\phi.$$

We have $\operatorname{curl}\mathbf{H} = \frac{1}{c}\operatorname{curl}\left[\frac{\partial}{\partial t}(\nabla\phi \times \mathbf{k})\right] = \frac{1}{c}\frac{\partial}{\partial t}[\operatorname{curl}(\nabla\phi \times \mathbf{k})]$.

By 7:5(14)

$$\nabla \times (\nabla\phi \times \mathbf{k}) = \nabla\phi(\nabla\,.\,\mathbf{k}) - \mathbf{k}\nabla\,.\,\nabla\phi + (\mathbf{k}\,.\,\nabla)\nabla\phi - (\nabla\phi\,.\,\nabla)\,\mathbf{k}$$

$$= -\,\mathbf{k}\nabla^2\phi + \frac{\partial}{\partial z}(\nabla\phi),$$

since

$$\nabla\,.\,\mathbf{k} = 0, \; \mathbf{k}\,.\,\nabla = \frac{\partial}{\partial z}.$$

So

$$\operatorname{curl}\mathbf{H} = \frac{1}{c}\frac{\partial}{\partial t}\left[-\,\mathbf{k}\nabla^2\phi + \frac{\partial}{\partial z}(\nabla\phi)\right]$$

$$= \frac{1}{c}\frac{\partial^2}{\partial z\,\partial t}(\nabla\phi) - \frac{1}{c}\frac{\partial}{\partial t}(\nabla^2\phi)\mathbf{k}. \tag{1}$$

Also

$$\frac{1}{c}\frac{\partial}{\partial t}\mathbf{E} = \frac{1}{c}\frac{\partial}{\partial t}\left[(\mathbf{k}\,.\,\nabla)\nabla\phi - \frac{1}{c^2}\mathbf{k}\frac{\partial^2\phi}{\partial t^2}\right]$$

$$= \frac{1}{c}\frac{\partial^2}{\partial z\,\partial t}(\nabla\phi) - \frac{1}{c^3}\frac{\partial}{\partial t}\left(\frac{\partial^2\phi}{\partial t^2}\right)\mathbf{k}. \tag{2}$$

From (1) and (2),

$$\operatorname{curl}\mathbf{H} = \frac{1}{c}\frac{\partial}{\partial t}\mathbf{E}$$

if

$$\frac{\partial^2\phi}{\partial t^2} = c^2\nabla^2\phi.$$

Similarly,

$$\begin{aligned}
\operatorname{curl}\mathbf{E} &= \operatorname{curl}\left[(\mathbf{k}\,.\,\nabla)\nabla\phi - \frac{1}{c^2}\frac{\partial^2\phi}{\partial t^2}\mathbf{k}\right] \\
&= \operatorname{curl}\left[\frac{\partial}{\partial z}(\nabla\phi)\right] - \frac{1}{c^2}\operatorname{curl}\left[\frac{\partial^2\phi}{\partial t^2}\mathbf{k}\right] \\
&= \operatorname{curl}\left[\frac{\partial^2\phi}{\partial z\,\partial x}\mathbf{i} + \frac{\partial^2\phi}{\partial z\,\partial y}\mathbf{j} + \frac{\partial^2\phi}{\partial z^2}\mathbf{k}\right] - \\
&\qquad - \frac{1}{c^2}\left[\frac{\partial^2\phi}{\partial t^2}\operatorname{curl}\mathbf{k} + \nabla\left(\frac{\partial^2\phi}{\partial t^2}\right)\times\mathbf{k}\right] \\
&= -\frac{1}{c^2}\nabla\left(\frac{\partial^2\phi}{\partial t^2}\right)\times\mathbf{k} \\
&= -\frac{1}{c^2}\frac{\partial}{\partial t}\left[\frac{\partial}{\partial t}(\nabla\phi\times\mathbf{k})\right], \text{ since } \nabla\frac{\partial^2\phi}{\partial t^2} = \frac{\partial^2}{\partial t^2}(\nabla\phi), \\
&= -\frac{1}{c}\frac{\partial\mathbf{H}}{\partial t}.
\end{aligned}$$

Thus **E** and **H** satisfy Maxwell's equations in free space provided

$$\frac{\partial^2\phi}{\partial t^2} = c^2\nabla^2\phi.$$

Exercises 8:6

1. Find the solid angle subtended by a thin hemispherical shell (shown in Fig. 67)

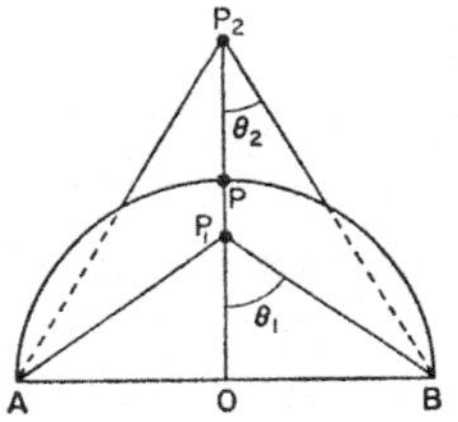

FIG. 67

(i) at its centre O, (ii) at P_1, where $\angle OP_1B = \theta_1$, (iii) at P_2, where $\angle OP_2B = \theta_2$.
Hence find the solid angle subtended at P on the surface as the limiting cases of (ii) and (iii) when P_1 and P_2 move up to P, and show that the solid angle changes by 4π when the hemispherical surface is crossed at P. [This discontinuity in the solid angle occurs whenever we pass from one side of a surface to the other, whether or not the surface is spherical.]

2. If ϕ is a scalar function of x, z and t, verify that the vectors

$$\mathbf{E} = \frac{\partial \phi}{\partial z}\mathbf{i} - \frac{\partial \phi}{\partial x}\mathbf{k}, \quad \mathbf{H} = -\frac{1}{c}\frac{\partial \phi}{\partial t}\mathbf{j}$$

satisfy Maxwell's equations in free space, provided that ϕ satisfies the equation

$$\frac{\partial^2 \phi}{\partial x^2} + \frac{\partial^2 \phi}{\partial z^2} = \frac{1}{c^2}\frac{\partial^2 \phi}{\partial t^2}.$$

3. The equations

$$\operatorname{curl} \mathbf{A} = \mu \mathbf{H}, \quad \operatorname{div} \mathbf{A} = -\frac{K\mu}{c}\frac{\partial \phi}{\partial t}, \quad \operatorname{grad} \phi = -\mathbf{E} - \frac{1}{c}\frac{\partial \mathbf{A}}{\partial t}$$

define the magnetic vector potential $\mathbf{A}$ and the electric scalar potential ϕ in the case where $\mathbf{E}$ and $\mathbf{H}$ both depend on t. Show that for a charge-free, non-conducting medium $\mathbf{A}$ and ϕ satisfy the equations

$$\nabla^2 \mathbf{A} = \frac{K\mu}{c^2}\frac{\partial^2 \mathbf{A}}{\partial t^2}, \quad \nabla^2 \phi = \frac{K\mu}{c^2}\frac{\partial^2 \phi}{dt^2}.$$

[In the case of steady currents, $\operatorname{div} \mathbf{A} = 0$ and $\mathbf{E} = -\operatorname{grad} \phi$; the above equations give the modifications necessary in the general case; ϕ, $\mathbf{A}$ are known as the *electromagnetic potentials.*]

4. Show that equations § 8:6 (iv) (15) and (16) are satisfied by the fields **E** and **H**, where

$$E_x = a\cos p\left(t - \frac{\sqrt{(K\mu)}}{c}z\right), \quad E_y = E_z = 0,$$

$$H_x = H_z = 0, \quad H_y = \sqrt{\left(\frac{K}{\mu}\right)}\, a\cos p\left(t - \frac{\sqrt{(K\mu)}}{c}z\right),$$

a, p being constants.
[This solution gives a plane wave of frequency $p/2\pi$ travelling in the direction of the z-axis with velocity $c/\sqrt{(K\mu)}$.]

5. Show that the potentials **A** and ϕ given by

$$A_x = A_y = 0, \quad A_z = -ax\sqrt{(K\mu)}\cos p\left(t - \frac{\sqrt{(K\mu)}}{c}z\right),$$

$$\phi = -ax\cos p\left(t - \frac{\sqrt{(K\mu)}}{c}z\right),$$

correspond to the plane wave in 4 above.

ANSWERS TO EXERCISES

Miscellaneous Exercises 1 (p. 5)

3. (i) $1/r^2$; (ii) $-2[(x-a)(y-b)+(y-b)(z-c)+(z-c)(x-a)]/r^4$, where $r^2 = (x-a)^2+(y-b)^2+(z-c)^2$.

4. (i) 0; (ii) $e^{2x}+e^{2y}+2e^{x+y}\cos(x+y)$;
(iii) $8e^{x+y}[\cos(x+y)+\sin(x+y)]$.

5. (i) 3; (ii) $5\mathbf{i}+7\mathbf{j}-13\mathbf{k}$; (iii) $3(\mathbf{i}-\mathbf{k})$; (iv) 12.

6. (i) $\sqrt{14}$; (ii) $3\sqrt{2}$; (iii) $9\sqrt{3}$;
$\hat{\mathbf{A}} = \mathbf{A}/\sqrt{14}$; $\hat{\mathbf{B}} = \mathbf{B}/3\sqrt{2}$; $\hat{\mathbf{C}} = (5\mathbf{i}+7\mathbf{j}-13\mathbf{k})/9\sqrt{3}$.

7. $\mathbf{A}\times\mathbf{B} = 5\mathbf{i}+22\mathbf{j}-19\mathbf{k}$. **8.** $\mathbf{x} = [p\mathbf{a}+(\mathbf{c}\times\mathbf{b})]/(\mathbf{c}\,.\,\mathbf{a})$.

12. $-\mathbf{i}-\mathbf{j}\pm3\mathbf{k}$.

Exercises 2:3 (p. 16)

3. $V = A\tan^{-1}(y/x)+B$; A, B arbitrary constants.

Exercises 2:4 (p. 21)

2. $\phi = (A\rho+B\rho^{-1})\sin\psi$; A, B arbitrary constants.

Exercises 2:5 (p. 24)

1. $df_1 = 3u^2\,du+3v^2\,dv+3x^2\,dx-3\,dy$;
$df_2 = 2u\,du+2v\,dv+2y\,dy+2dx$;

$$\frac{\partial u}{\partial x} = \frac{v-x^2}{u(u-v)};\quad \frac{\partial u}{\partial y} = \frac{1+vy}{u(u-v)};\quad \frac{\partial v}{\partial x} = \frac{u-x^2}{v(v-u)};\quad \frac{\partial v}{\partial y} = \frac{1+uy}{v(v-u)}.$$

3. $$\frac{\partial x}{\partial u} = \frac{u^3(9u^2-8y^2)}{6x^2(x-y^2)}.$$

Exercises 2:6 (p. 28)

2. (i) $\dfrac{\partial(x,y)}{\partial(\rho,\psi)} = \rho$; (ii) $\dfrac{\partial(x,y,z)}{\partial(r,\theta,\psi)} = r^2\sin\theta$;

(iii) $\dfrac{\partial(x,y,z)}{\partial(u,v,w)} = u^2+v^2$.

Miscellaneous Exercises 2 (p. 28)

3. (i) $z = f[1/(x+y)] - xy(x+y)$; (ii) $z = f[(x+y)/xy] - xy$; f an arbitrary function in each case.
4. $\phi = (x^2+y^2)^2 \log 2xy - 2x^2y^2 + f(x^2+y^2)$; f an arbitrary function.

15. (i) $\dfrac{d^2F}{\partial x^2} + \dfrac{\partial^2 F}{\partial y^2} = \dfrac{\partial^2\phi}{\partial\rho^2} + \dfrac{1}{\rho^2}\dfrac{\partial^2\phi}{\partial w^2} + \dfrac{1}{\rho}\dfrac{\partial\phi}{\partial\rho}$;
(ii) $\psi = A\tan^{-1}(e^u) + B$; A, B arbitrary constants.
18. $F = A\log xy + B$; A, B arbitrary constants.

Exercises 3:1 (p. 38)

1. (i) $23a^4/2$; (ii) $10a^4$; (iii) $86a^4/7$.
2. $a^2(3b\pi + 4a)/3$. **3.** 0.

Exercises 3:2 (p. 42)

1. $\pi/4 - \tan^{-1}(2/3) + \log(13/2)$.
2. $\alpha = -1$; (i) $(3+\alpha)/2$; (ii) $(1+\alpha)\pi/4 + (1-\alpha)/2$.

Exercises 3:3 (p. 45)

1. $\frac{1}{2}\log(2+\sqrt{3})$. **2.** $2\pi/3\sqrt{a}$.

Exercises 3:4 (p. 47)

1. (i) 1; (ii) 1; (iii) 36/35. **2.** 6.

Miscellaneous Exercises 3 (p. 48)

1. -13. **3.** (i) 1/3; (ii) $(15\pi^3 - 8\pi^2 + 36\pi - 192)/24\pi^2$.
4. 9. **5.** $a = 2$; $b = 1$; $(7\pi - 2 - 2e^8)/2$.
6. 888 *units*. **7.** (i) $3a^3/2$; (ii) $3\pi a^2/4$.
8. (i) $(\sqrt{2}/a^2)\sinh 8\pi$; (ii) (a) $2\sqrt{2}\pi a^3(3-4\pi^2)/3$.; (b) $2a\pi(1+6a^2)$.

Exercises 4:2 (p. 57)

2. 48/7. **3.** $-2/\pi^2$; $2/\pi^2$. **4.** $(3a^2\log 3a)/8$.
5. $(4\pi - \pi^2 - 8)/4\pi$. **6.** $\pi/2$. **7.** 1/6.

Exercises 4:3 (p. 59)

1. $(e-2)/2$. **2.** 4/9. **3.** 1/4. **4.** $(a^2\sinh^{-1}1)/2$.

Exercises 4:4 (p. 64)

1. $(8\log 2 - 5)/4a$. **3.** 46/81. **4.** $(e^2-1)/4e$.

Exercises 4:5 (p. 68)

2. $(2\pi + 3\sqrt{3})/9$. **3.** $3a^2[\pi + 12\tan^{-1}(1/3)]/64$.

Exercises 4:6 (p. 72)

1. 0. **2.** -100π. **3.** $(4\pi - 8 - \pi^2)/4\pi$.

Exercises 4:7 (p. 77)

1. $a^6/18$. **2.** $1/5$. **3.** $2/15$.

Exercises 4:8 (p. 88)

2. $4\pi[a^3 - (a^2 - b^2)^{3/2}]/3$.
3. $I_{OX} = M(b^2 + c^2)/5$; $I_{OY} = M(c^2 + a^2)/5$; $I_{OZ} = M(a^2 + b^2)/5$; where M is the mass of the solid.
4. $\pi a^3/2$; $(0, 0, a/3)$. **5.** $k_x = a\sqrt{7/22}$; $k_y = a\sqrt{7/3}$.

Miscellaneous Exercises 4 (p. 88)

1. $a^4/12$. **2.** $\pi a^4 b^2/5$. **4.** $181\pi/216$.
5. Change of order gives $\displaystyle\int_0^{4a} \frac{dx}{4a - x} \int_x^{2\sqrt{(ax)}} \frac{y\,dy}{\sqrt{(4ax - y^2)}}$; $2a\pi$.
6. $[(a^2 + 1)^{3/2} - (b^2 + 1)^{3/2} - a^3 + b^3]/3$, assuming $a > b$.
7. $\pi abcm/(m + 1)$. **9.** $2ar^3/3c$.
10. $\displaystyle V = \int_0^{\pi/2} \sin^2\theta\,d\theta \int_0^{\infty} r^3 e^{-r^2}\,dr = \pi/8$.
12. $137a^3/20$. **13.** $13\pi a^3/24$. **15.** $\pi R^2(2a^2 - 3R^2)/2$.
18. $(3a/8, 3b/8, 3c/8)$. **20.** $a^4[3\log(\sqrt{2} + 1) - \sqrt{2}]/32\sqrt{2}$.
22. $a(3\pi - 2)/[32(2 - \sqrt{2})]$. **23.** $2(3\pi - 10)/3$.
24. $\pi\sqrt{2}/12$.

Exercises 5:2 (p. 96)

1. $8\pi^2$. **3.** Surface is given by $0 \leqq \theta < \pi$, $0 \leqq \phi < 2\pi$.
4. $\pi(37\sqrt{37} - 1)/6$.

Exercises 5:4 (p. 101)

1. $\pi a^4/4$. **2.** 64π. **3.** $12\pi a^6/5$. **4.** $4\pi(a^2 + 2b^2)/3b$.

Exercises 5:5 (p. 107)

1. $24a^5$. **3.** $6\pi a^3$. **4.** $a^3(9\pi a + 32)/48$.

Miscellaneous Exercises 5 (p. 108)

1. $I = h\pi^2[a\sqrt{(a^2+h^2)} + h^2\log\{a+\sqrt{(a^2+h^2)}\} - h^2\log h]/4$.
2. $2(\pi + 4 - 4\sqrt{2})$.
3. Centre inside cylinder: $2b\int_0^{\pi} d\theta \int_{r_1}^{r_2} [r/\sqrt{(b^2-r^2)}]\, dr$, where

   $r_1 = c\cos\theta - \sqrt{(a^2 - c^2\sin^2\theta)}$, $r_2 = c\cos\theta + \sqrt{(a^2 - c^2\sin^2\theta)}$;

   Centre outside cylinder: $4b\int_0^{\alpha} d\theta \int_{r_1}^{r_2} [r/\sqrt{(b^2-r^2)}]\, dr$, where

   $\sin\alpha = a/c$;

   (i) $4b\pi[b - \sqrt{(b^2-a^2)}]$; (ii) $2b^2(\pi - 2)$.
6. $16384\sqrt{2}/693$. **7.** $\sqrt{2}\pi a^3/3$. **8.** $-256\pi/3$.
9. $2\pi a^5/3$. **10.** $a^2(3\pi + 8a)/12$. **11.** $-5a^3$; 0.

Exercises 6:2 (p. 117)

1. $2\mathbf{i} + 2\mathbf{j} + 2\sqrt{2}\mathbf{k}$; 4; $\frac{1}{2}\mathbf{i} + \frac{1}{2}\mathbf{j} + \frac{1}{\sqrt{2}}\mathbf{k}$.
2. grad $\phi = 5\mathbf{i} + 2\mathbf{j} - 2\mathbf{k}$; $3\mathbf{i} + 2\mathbf{j} - 6\mathbf{k}$; 7.
3. The normals are $12\mathbf{i} + 8\mathbf{j} - 6\mathbf{k}$, $4\mathbf{i} - 2\mathbf{j} - \mathbf{k}$.

Exercises 6:3 (p. 120)

1. (i) $2\mathbf{r}$; (ii) $r^2(4 + 7r^2)\mathbf{r}$; (iii) $r^{-3}[r^2f'(r) - g'(1/r)]\mathbf{r}$.
2. (i) $\mathbf{a}$; (ii) $f(r)\mathbf{a} + f'(r)(\mathbf{a}\,.\,\mathbf{r})\mathbf{r}/r$; (iii) $(\mathbf{a}\,.\,\mathbf{r})\mathbf{b} + (\mathbf{b}\,.\,\mathbf{r})\mathbf{a}$.

Exercises 6:4 (p. 122)

1. 402. **2.** $12\pi\mathbf{k}$.

Exercises 6:5 (p. 126)

1. $\phi = x^3y + xy^2z +$ constant; 402.
2. (i) and (iv), irrotational; (ii) and (iii), rotational;
   In (i), $\phi = r^4/4 +$ constant.

Exercises 6:7 (p. 131)

1. (i) $3x^2y + 3y^2z + 2z$; (ii) $2(2x^2 + 2y^2 + z^2 - z)$.
2. (i) $3f(r) + rf'(r)$; (ii) $4(\mathbf{a}\,.\,\mathbf{r})$; (iii) $(3+n)r^nf(1/r) - r^{n-1}f'(1/r)$;
   (iv) 0; (v) 0; (vi) $y + x$.

Exercises 6:8 (p. 133)

1. (i) $(n+3)r^n$; (ii) $4(\mathbf{a}\,.\,\mathbf{r})$; (iii) $8r$; (iv) $10\mathbf{r}$.

Exercises 6:9 (p. 136)

2. No; irrotational only.

Exercises 6:10 (p. 142)

1. 81π; -81π; 162π.
2. (i) 200π; (ii) $50(20+3\pi)/3$.
3. $\dfrac{4\pi}{15}\left[\dfrac{3bc}{a^3}+\dfrac{3ca}{b^3}+\dfrac{3ab}{c^3}+\dfrac{2a}{bc}+\dfrac{2b}{ca}+\dfrac{2c}{ab}\right]$.
4. $ab^2c^2(2a^2-c^2)/4$.

Miscellaneous Exercises 6 (p. 142)

2. (i) 6; (ii) 6. **3.** $\phi = y^2z^3\sin x - x^4z + c$.
4. 35. **5.** $\operatorname{div}\operatorname{grad}\phi = 0$.
8. (i) $18r^{-1}(\mathbf{a}\,.\,\mathbf{r})(\mathbf{b}\,.\,\mathbf{r})$; (ii) $6r(\mathbf{a}\,.\,\mathbf{r}) + 12r^{-5}(\mathbf{b}\,.\,\mathbf{r})$; (iii) 0.
11. $16\pi abc/3$. **12.** $8\pi\mathbf{k}$. **14.** $(3a/8, 3a/8, 3a/8)$.
16. $n(n+2)(n+4)(n-2m)(n+2-2m)(n+4-2m)r^{-(n+6)}V$.
18. (ii) $\displaystyle\int_{S'}\frac{\partial}{\partial n}\left(\frac{1}{r}\right)\mathrm{d}S \to -4\pi$ as $\rho \to 0$.

Exercises 7:1 (p. 151)

1. $2x(y-2)\mathbf{i} - y^2\mathbf{j} + (4z-x^2)\mathbf{k}$; $-2\mathbf{i}-\mathbf{j}+3\mathbf{k}$; $-4\mathbf{j}+11\mathbf{k}$.
2. $z(9z-14x)\mathbf{i} + 8yz\mathbf{j} + (3z^2-4x^2)\mathbf{k}$.
3. (i) $\mathbf{0}$; (ii) 0. **4.** (i) $\mathbf{0}$; (ii) $\mathbf{0}$.

Exercises 7:2 (p. 155)

3. (i) $-2\mathbf{A}$; (ii) $-2x(y+z)\mathbf{i} - 2y(z+x)\mathbf{j} - 2z(x+y)\mathbf{k}$.

Exercises 7:3 (p. 158)

1. $m = 0$ or $n = -1$.
2. (i) rotational; (ii) and (iii) irrotational.

Exercises 7:4 (p. 162)

1. $3\pi/2$. **2.** $-\pi$. **4.** $6\pi a^3$.

Exercises 7:5 (p. 170)

1. (i) 0; (ii) $20r^2(\mathbf{a}\,.\,\mathbf{b})$; (iii) $-9r^{-4}[(\mathbf{a}\,.\,\mathbf{r})(\mathbf{b}\times\mathbf{r}) + (\mathbf{b}\,.\,\mathbf{r})(\mathbf{a}\times\mathbf{r})]$.

Exercises 7:6 (p. 175)

1. (i) $dl^2 = dr^2 + r^2 d\theta^2 + r^2 \sin^2\theta \, d\psi^2$; $dV = r^2 \sin\theta \, dr \, d\theta \, d\psi$;
(ii) $dl^2 = d\rho^2 + \rho^2 d\psi^2 + dz^2$; $dV = \rho \, d\rho \, d\psi \, dz$;
(iii) $dl^2 = (u^2 + v^2)(du^2 + dv^2) + u^2 v^2 d\psi^2$; $dV = uv(u^2 + v^2) \, du \, dv \, d\psi$.

Exercises 7:8 (p. 191)

1. The three vector normals at the point (x_1, y_1, z_1) are
$x_1\mathbf{i} + y_1\mathbf{j} + z_1\mathbf{k}$, $x_1\mathbf{i} + y_1\mathbf{j} - c_2^2 z_1\mathbf{k}$, $-c_3\mathbf{i} + \mathbf{j}$.
2. Components of $\nabla\phi$ are $2\xi/(\xi^2 + \eta^2)^{3/2}$, $2\eta/(\xi^2 + \eta^2)^{3/2}$, 0.

$$\nabla^2 f = \frac{1}{(\xi^2 + \eta^2)}\left[\frac{\partial^2 f}{\partial \xi^2} + \frac{\partial^2 f}{\partial \eta^2}\right] + \frac{1}{\xi^2\eta^2}\frac{\partial^2 f}{\partial \psi^2} + \frac{1}{(\xi^2 + \eta^2)}\left[\frac{1}{\xi}\frac{\partial f}{\partial \xi} + \frac{1}{\eta}\frac{\partial f}{\partial \eta}\right].$$

3. (i) Coordinate surfaces are
(a) elliptic cylinders, $\dfrac{x^2}{k^2\cosh^2 c_1} + \dfrac{y^2}{k^2\sinh^2 c_1} = 1$, generators parallel to the z-axis, given by $u = c_1$;
(b) hyperbolic cylinders, $\dfrac{x^2}{k^2\cos^2 c_2} - \dfrac{y^2}{k^2\sin^2 c_2} = 1$, generators parallel to the z-axis, given by $v = c_2$;
(c) planes parallel to $z = 0$, given by $z = c_3$.
(ii) $dl^2 = k^2(\sinh^2 u + \sin^2 v)(du^2 + dv^2) + dz^2$.
$dV = k^2(\sinh^2 u + \sin^2 v)\, du \, dv \, dz$,

$$\nabla\phi = \frac{1}{k(\sinh^2 u + \sin^2 v)^{1/2}}\left(\frac{\partial\phi}{\partial u}\hat{\mathbf{u}} + \frac{\partial\phi}{\partial v}\hat{\mathbf{v}}\right) + \frac{\partial\phi}{\partial z}\hat{\mathbf{z}},$$

$$\nabla^2\phi = \frac{1}{k^2(\sinh^2 u + \sin^2 v)}\left[\frac{\partial^2\phi}{\partial u^2} + \frac{\partial^2\phi}{\partial v^2} + k^2(\sinh^2 u + \sin^2 v)\frac{\partial^2\phi}{\partial z^2}\right].$$

(iii) $\hat{\mathbf{u}} = (\sinh^2 u + \sin^2 v)^{-1/2}(\sinh u \cos v\mathbf{i} + \cosh u \sin v\mathbf{j})$,
$\hat{\mathbf{v}} = (\sinh^2 u + \sin^2 v)^{-1/2}(-\cosh u \sin v\mathbf{i} + \sinh u \cos v\mathbf{j})$,
$\hat{\mathbf{z}} = \mathbf{k}$.

4. Coordinate surfaces are
(a) parabolic cylinders $y^2 = c_1^2(c_1^2 - 2x)$, generators parallel to the z-axis, given by $u = c_1$;
(b) parabolic cylinders $y^2 = c_2^2(2x + c_2^2)$, given by $v = c_2$;
(c) planes parallel to $z = 0$, given by $z = c_3$.
The unit vectors are
$\hat{\mathbf{u}} = (u\mathbf{i} + v\mathbf{j})/\sqrt{(u^2 + v^2)}$, $\hat{\mathbf{v}} = (-v\mathbf{i} + u\mathbf{j})/\sqrt{(u^2 + v^2)}$, $\hat{\mathbf{z}} = \mathbf{k}$.

Miscellaneous Exercises 7 (p. 192)

1. (i) $\operatorname{div}(\mathbf{A}\times\mathbf{B}) = 0$, $\operatorname{curl}(\mathbf{A}\times\mathbf{B}) = -5x\mathbf{i} + 10y\mathbf{j} - 5z\mathbf{k}$.

4. (i) $A_1 = \dfrac{x}{3r^3} + \dfrac{\partial\phi}{\partial x}$, $A_2 = -\dfrac{y}{3r^3} + \dfrac{\partial\phi}{\partial y}$, ϕ an arbitrary function of x and y.

5. $\pi\mathbf{k}/a\sqrt{2}$. **6.** $2\pi ab(2\mathbf{i}+\mathbf{j})$. **12.** $Ar + Br^{-2}$.

13. $$\nabla^2\phi = \frac{\partial^2\phi}{\partial\lambda^2} + \frac{\partial^2\phi}{\partial\mu^2} + (\lambda^2+\mu^2)\frac{\partial^2\phi}{\partial z^2};$$
$$\frac{d^2M}{d\mu^2} - (a^2\mu^2 - b)M = 0.$$

14. u = const. gives the surface of revolution $z = c(x^2+y^2)^2$, obtained by rotating the curve $z = cy^4$ in the y–z plane about the z-axis;
v = const. gives the ellipsoids $4z^2 + x^2 + y^2 = c$;
ϕ = const. gives the planes $y = cx$.
Also, $h_u^2 = \mu^8/H$, $h_v^2 = 1/(4\mu^2H)$, $h_\phi = \mu$.

15. $$\nabla\phi = \frac{\partial\phi}{\partial\rho}\hat{\boldsymbol{\rho}} + \frac{1}{\rho}\frac{\partial\phi}{\partial\psi}\hat{\boldsymbol{\psi}} + \frac{\partial\phi}{\partial z}\hat{\mathbf{z}};$$
$$\nabla\cdot\mathbf{A} = \frac{1}{\rho}\frac{\partial}{\partial\rho}(\rho A_1) + \frac{1}{\rho}\frac{\partial A_2}{\partial\psi} + \frac{\partial A_3}{\partial z};$$
$$\nabla^2\phi = \frac{1}{\rho}\frac{\partial}{\partial\rho}\left(\rho\frac{\partial\phi}{\partial\rho}\right) + \frac{1}{\rho^2}\frac{\partial^2\phi}{\partial\psi^2} + \frac{\partial^2\phi}{\partial z^2}.$$

Exercises 8:6 (p. 253)

1. (i) 2π; (ii) $2\pi(1+\cos\theta_1)$; (iii) $2\pi(\cos\theta_2 - 1)$; $2\pi(1 + 1/\sqrt{2})$; $2\pi(1/\sqrt{2} - 1)$.

This page intentionally left blank

INDEX

(The numbers refer to pages. Figures in italics indicate that the reference is to an exercise on that page.)

Commonwealth Library "Higher Mathematics for Scientists and Engineers"

Advanced Calculus and Vector Field Theory, Kathleen M. Urwin.
Differential Equations and Special Functions, F. M. Arscott.
Complex Variable and Integral Transforms, H. V. Miller.
Advanced Dynamics, W. Davidson.
Numerical Analysis, I. M. Khabaza.